Anonymous

The Art of Paper-Making

A guide to the theory and practice of the manufacture of paper: being a

compilation from the best known French, German and American writers

Anonymous

The Art of Paper-Making
A guide to the theory and practice of the manufacture of paper: being a compilation from the best known French, German and American writers

ISBN/EAN: 9783337222284

Printed in Europe, USA, Canada, Australia, Japan

Cover: Foto ©berggeist007 / pixelio.de

More available books at **www.hansebooks.com**

THE ART

OF

PAPER-MAKING:

A Guide to the Theory and Practice

OF THE

MANUFACTURE OF PAPER.

BEING A COMPILATION FROM THE BEST KNOWN
FRENCH, GERMAN, AND AMERICAN
WRITERS.

BY THE

EDITOR OF "THE PAPER MILLS DIRECTORY."

London:
SOLD BY KENT & CO., PATERNOSTER ROW.
1874.

Price 25*s.*

UNWIN BROTHERS,
PRINTERS.

PREFACE.

It is very remarkable that since the Art of Paper-making first struck root in this country, now more than three hundred years since, no work of any kind has appeared on the subject, of any authority or otherwise, of a technical character relating to the manufacturing process strictly; and adapted to the wants of the Paper-maker.

The French Paper-making trade and the German have—although only within the last ten years—produced some writers on the subject, whose writings and treatises have obtained authority to some extent among them; and quite recently a German writer in America has written a treatise of considerable practical merit. And the simple matter of making selections from these for the benefit of the English makers, is all we lay claim to in these pages. In the early part of the work the former have been drawn upon, and for the later portion the latter.

Hence it is that some parts of this work are, we are aware, either crude, obsolete, or useless, in comparison with our most modern modes of working. But it will at least serve the purpose in laying a foundation for future efforts, of showing what our foreign neighbours, whom we are so frequently threatened are to be our future formidable competitors, have adopted, and do still use, as text-books of their modes of operation.

London, 1874.

CONTENTS.

CHAPTER I.
A Sketch of the History of Paper and Paper-making Invention 1

CHAPTER II.
Raw Materials Employed in Paper-making 5

CHAPTER III.
Manufacturing Process Machine-made Paper 11

CHAPTER IV.
Manufacturing Process Machine-made Paper 66

CHAPTER V.
Manufacturing Process Hand-made Paper 72

CHAPTER VI.
Chemical Analysis of Paper-making Materials 94

CHAPTER VII.
Washing Water, Water Power, Steam Boilers, etc... 122

INDEX.

	NO.		NO.
Air Roll	106	Dyes	19
Alkali Test	35	Engine Power	68
Alums	38	,,	18
Ancient Period	1	Engines, Steam	65
Antichlorine (page)	31	,,	67
,,	37	Esparto	5
Apron	74	,, Paper	23
,,	84	Examination of Papers	48
Bar Screens	76	Expansion, Steam	65
Beating	13	Explosions, Boiler	64
,,	17	Fan Pump	73
Bleaching	15	Feed Water	63
Boiler Tests	62	Felts	101
,, Explosions	64	Felt Rolls	102
Boilers	57	,, Washer	105
,,	62	,, Management	108
Boiling Rags	12	Finishing	22
Brass Rolls	99	,,	32
Calenders	113	Firing	61
Chlorine	15	Foreign Rags (page)	13
Chlorometric tests	36	Fourdrinier Machine	71
Chemical Impurities	47	Fuel	42
Chilled Rolls	114	,, Combustion	59
Chimney Draught	60	,, for Drying	112
Clutch	107	Gates	95
Colouring	19	Gelatine	29
,,	20	Glazing	22
,, Pulp (page)	47	Grate Surface	61
,, Matters	41	Half Stuff	13
Combustion and Fuel	59	Hand-made Papers	26
Condensation	66	Heating Surface	58
Construction of Boilers	62	Housing	98
Couch Rolls	86	Ibotson's Strainer	78
Cutters	118	Kaolin	39
Dandy Roll	80	Laboratory	44
Deckles	94	Limes	36
Doctors	100	Machine	21
Draining	14	,, Wire	85
Draught Chimney	60	Manganese	36
Drying	31	Middle Ages Period	2
,,	111	Mill Dams	53
Drying Cylinders	110	Mixing Box	73
Dusting	10	Modern Period	3

	NO.
Motive Power	120
Overshot Wheels	56
Paper, Examination of	43
Paper in Rolls	119
,, Carrying Rolls	102
,, Machine	21
,, ,,	71
,, Shavings	8
Plate Screens	77
Power	56
,, Water	52
,, Engine	68
,, Motive	120
,, Loss of	69
Press Felt	103
,, Rolls	98
Pulp	16
,, Diluting	72
,, Colouring	20
,, ,, (page)	47
,, Dresser	74
,, Regulator	72
Pumps	51
Rag Papers (page)	11
,, Boiling	12
,, Dusting	10
,, Washing	11
Rags	4
,, Foreign (page)	13
Reels	116
Refining	17
Reversed Screens	81
Revolving ,,	80
Rolls, Air	106
,, Brass	99
,, Chilled	114
,, Felt	102
,, Paper Carrying	102
,, Rubber Cased	99
,, Spread	104
,, Stretch	104
Rubber Cased Rolls	99
Sand Tables	74
Save-all	90
Screen Vat	83
Screens, see Bar Screens.	
,, ,, Plate do.	
,, ,, Revolving do.	
,, ,, Reversed do.	
Shaking motion	93

	NO.
Size and Speed	121
Sizing	18
,,	27
,, Room	28
Sorting and Cutting	9
Spread Roll	104
Starch	40
Steam Boilers	57
,, Engines	65
,, ,,	70
,, Utilisation	67
Steaming Paper	115
Slitters	117
Strainers	75
,,	78
,,	82
,, Suction	79
Straw	6
Straw Papers	24
Stretch Rolls	91
,,	104
Stuff	16
,, Catchers	92
Suction Boxes	88
Tinting	19
Trimming and Cutting	117
Tube Rolls	87
Turbines	55
,,	56
Ultramarine (page)	45
Utilisation of Steam	67
Water Distribution	49
Water-marking	33
Waters	34
Washing Rags	11
,, Water	45
,,	50
Waste Paper	8
Water Pipes	90
,, Power	52
,, Quantity	50
,, Wheels	54
Wood	7
,, Paper	25
Webs and Paper	119
Wet Felt	103
Wire Cloth	85
,, ,,	96
Wire Guides	97

THE ART

OF

PAPER-MAKING:

A Guide to the Theory and Practice

OF THE

MANUFACTURE OF PAPER.

COMPILED BY THE

EDITOR OF THE "PAPER MILLS DIRECTORY."

London:
SOLD BY UNWIN BROTHERS,
24, BUCKLERSBURY.

B. DONKIN & CO.'S SPHERICAL ROTARY RAG BOILER.

TO CONTAIN FROM 20 TO 25 CWT. OF RAGS.

This Boiler being spherical, is twice as strong as a Cylindrical one of the same diameter and thickness. The Plates used are notwithstanding of the usual substance, thus rendering it perfectly safe, durable, and suitable for high pressure steam.

The Spherical shape has also another important advantage, viz. allowing all the rags to fall out by themselves, when the boiler is revolving with its cover off.

Inside the Boiler are Strainers to take off the dirt, and Lifters to agitate the rags during the process of either boiling or washing.

To avoid Cement or run lead Joints, the Gudgeons and the Boiler are turned true in a lathe to fit each other, the joints being simply made with red lead.

B. DONKIN & CO.,
ENGINEERS, MILLWRIGHTS, MACHINISTS, AND IRON FOUNDERS,
Blue Anchor Road, Bermondsey, London, S.E.

Manufacturers of Steam Engines, Turbines, Water Wheels, Flour Mills, &c., Paper Making Machines, and everything connected with Paper Mills, Rag Engines, Rag Cutters, Cutting Machines, Glazing and Hydraulic Presses, Pumps, Hoists, &c., Steel Bars and Plates, Wires, Felts, and Deckle Straps. Machines for Printing in two Colours. Gas and Water Valves.

are crushed, and appear as transparent yellow spots when the leaf is held up to the light.

The work of a satining apparatus requires a number of hands; three or four gangs besides one or two workmen to hand the plates. One will be sufficient, if the rollers work in both directions so as to return when reversed.

A boy or girl may place and remove the sheets of paper, and a workman lays on the zinc plate. Sometimes this work is confided to a woman, but a man is indispensable for the larger sizes.

Calenders, or three cylindrical presses, are sometimes used, the middle one being of paper and the other two of metal. This apparatus is absolutely necessary to glaze paper in rolls. The paper is wound upon rollers placed at the head of the calender, and passes first between the upper and middle cylinder, and then between the middle and lower ones.

This finishing in rolls, especially when the paper is very wide, presents some difficulties. It is necessary that there should be a uniform pressure upon every point of contact, otherwise the paper will be creased and a considerable loss occur.

This accident is less to be apprehended when the paper is rolled in separate sheets.

Three hands are necessary to attend to the work of a calender. The first engages the leaf between the two upper cylinders; the second returns it between the middle and the lower; and the third receives the sheet.

The calendered paper preserves its whiteness, which is somewhat impaired by zinc. All things being equal, the plates are better for thin papers.

The glazing done, there remains the operation of counting and putting up into reams.

The sheets are counted by special hands, who acquire great skill.

The quires, after being arranged in reams, with the backs placed alternately for folded papers, are put under the press at night, and the next day are ready to be packed.

Bundles are composed of two reams.

It is well to weigh before packing; and to label each ream, indicating the quantity, quality, &c.

A careful manufacturer should see that the packing is always done with care, and even with elegance, in the case of fine papers.

Chapter IV.

MANUFACTURE.

23. *Esparto Paper.*—There are found in Esparto two or three qualities as imported for Paper-making, all of which should be sorted and cleaned by hand before they are suitable for putting into the engine. This may be done on the ordinary rag-cutting tables.

Weeds, roots, or flowers should be carefully cut off and thrown out. It should, like straw, be washed and boiled with caustic alkali, washed and bleached with a chlorine solution.

Before subjecting this material to the boiling and bleaching processes, it is essential to crush it in the direction of its length, by means of fluted rollers; as the fibres are more permeable to chemical agents when separated.

It may be put into the boiler in the dry state and in its full length.

Esparto, after being boiled with lye, retains enough tenacity to be separated into long shreds. It contains a red colouring matter which becomes soluble under the combined influence of chlorine and caustics.

The waters used in washing assume a blood-red tinge.

The waste of material may be thus estimated :—

Yellow colouring matters	12
Red ,, ,,	6
Resin and gum	7
Salts constituting the ash	1·500
	26·500
Paper fibres	73·500
	100·000

The yellow colouring matter being soluble in alkalies, is eliminated after the first boiling.

By mixing Esparto with rags of poorer quality, very good printing paper may be produced.

Stationary boilers are considered to be best adapted for this material.

In the preparation of the lye it is essential that the soda should be rendered caustic.

The yield of stuff from Esparto is considered to be not less than 40 or more than 50 per cent.; the variation being caused by the varying purity and quality of the grass itself.

24. **Straw Papers.**—The first attempts to make paper from straw go back to the beginning of this century. The process consisted in subjecting the material to the action of a lye made of a mixture of lime and soda, or potash, until the substance was softened enough to be crushed between the fingers. After washing followed by trituration, the stuff was converted into paper.

The numerous patents taken out with the same object are based upon analogous principles. Packing paper of considerable strength has been made by mixing 50 per cent. of coarse rags with straw pulp, prepared in the following manner:—

220 lbs. of wheat straw, finely chopped, and 176 lbs. of quick-lime, were placed in a boiler with a sufficient amount of water to form a kind of pulp. The mixture was stirred, and poured into a second boiler every day for a fortnight.

This material was then reduced to pulp, and mixed with the rag pulp in the beating engine. The product obtained was half-sized, of a yellowish tinge and great strength.

Straw contains a yellow colouring matter, which is more or less communicated to the paper, unless subjected to a succession of boilings and bleachings, with gaseous or liquid chlorine. In this case the waste of material is very much increased.

The majority of manufacturers who employ straw are satisfied with subjecting it to maceration with lime, and succeed in making common wrapping paper. If, however, it is wished to make common printing paper, such as newspaper, &c., of straw pulp, it is indispensable to bleach with repeated chlorine and acid baths. The mixture is then made in the proportion of—

    25 to 40 per cent. of straw pulp.
    75 ,, 60   ,,   ,, rag pulp.

The manufacture is thus only profitable in localities where chemicals are at a low price. Very fair papers have been made with 80 to 90 per cent. of straw, but it remains to be seen whether the profits will render the process practicable.

The nature of the straw, and the composition of the soils in which it is grown, are not unimportant matters to the paper-maker. The hardness of straws proceeds from the quantity of silica interspersed throughout their substance, forming an obstacle to their conversion into paper, by binding together the fibrous parts of the stalk.

The straw to be chosen is that of wheat, as being the most tender. This fact is in accordance with chemical analysis, which gives the following proportions of silica contained in the three principal cereals of our soil:—

    1st. Wheat straw, 4·3 per cent. of silica.
    2nd. Rye   ,,   6·3 ,,   ,,
    3rd. Barley ,,   6·9 ,,   ,,

The knots of graminious plants, in general, are injected with a much greater quantity of silica than the intermediate parts, and should therefore be carefully removed when the straw is to be converted into white paper pulp.

Among other straws which have been tried, we may mention maize, which gives a naturally sized paper of great strength, and which at one time engaged some attention.

Maize stalks unbleached, only boiled with lye, and added to rag pulp, are suitable for making packing-paper. The product possesses a certain tenacity not offered by that of other straws. This substance may then answer very well in countries where it grows in sufficient quantities to allow of its employment.

25. Wood Paper.—For twenty years wood has been the object of numerous, and at first fruitless, experiments. At the present day, owing to the persevering efforts of a few inventors, processes are known which admit of the use of this material, when mixed with rags, for making the commoner kinds of printing paper.

In France and Belgium, white and resinous woods, such as pine, fir, &c., are employed.

It has been shown by experiments that wood is not made up of cells solidified by encrusting materials, as was long supposed, but of two superimposed layers; one exterior, short-grained, and brittle, the other internal, supple, and fibrous.

In order to make wood paper, all that is necessary is to dissolve the former and employ the remaining substance.

This first layer is acted upon by alkalies, hydrochloric acid, and the hypochlorites.

The machines employed to separate the fibres of wood are of different kinds. That invented by M. Vœlter consists in a grindstone, which consumes several logs of wood placed against its circumference and pressed in such a manner as to keep them always in contact with it during the entire duration of the work.

Another machine crushes the chips between cast iron rollers, and the material is triturated by means of a cylinder engine. The pulp is strained through a sieve, placed in the cap of the engine, and may then be used immediately for the manufacture of common papers. Wood alone gives too brittle a paper, and the short thick fibres do not yield that suppleness which is characteristic of rag paper.

The motive force necessary to separate the fibres of wood is very great. It is calculated that a 25 horse power is required for 24 hours, in order to produce 6 cwt. of wood pulp.

We give as an example of the mixture the proportions admitted by M. Vœlter, of Heidenheim.

	Per cent.		Per cent.	
Writing paper	33 of poplar wood,	67 of rags.		
Fine printing paper . .	20	,,	,,	80 ,,
Common printing paper .	50 of pine wood	50	,,	
Brown wrapping ,, .	40	,,	,,	60 ,,
Grey ,, ,, .	50	,,	,,	50 ,,
Blue paper	33	,,	,,	67 ,,

In Belgium, paper for newspapers is made of—
  60 parts birch wood,
  20 ,, kaolin,
  20 ,, coarse grey rags.

The wood pulp is added a quarter of an hour before the end of the beating process. The wood being neither boiled nor bleached, the proportion of bleaching should be diminished, as a part of the colouring matter is not destroyed by acids.

The use of chemical agents to bleach wood pulp requires much skill, otherwise there is great risk of having a pulp without consistency, giving rise to considerable loss of material.

The introduction of unboiled and unbleached pulp is considered so great an economy, that several manufacturers of Germany, Belgium, and France have decided to make additions to their paper-mills, in order specially to prepare wood pulp.

Chapter V.

MANUFACTURE.

26. Hand Made Paper.—Paper-making from the vat, or by hand, requires the labour of three workmen for each vat: 1st. The vatman; 2nd. The coucher; 3rd. The layman.

The first, by dipping the mould into the vat, makes the sheet; the second stretches or couches it upon the felt; the third, when the post has been pressed, successively detaches each damp sheet interposed between the felts.

The workman, holding the mould in both hands by the two short sides, dips it into the pulp at an inclination of about 60 or 70 degrees, and raises it horizontally, after having taken up enough of the pulp to obtain the required thickness of the paper he intends to make.

By means of a double oscillating motion, called balancing, he distributes the pulp, as uniformly as possible, over the entire surface of the mould. Gradually the water drains through the wires or the meshes of the wire gauze; the pulp solidifies, and assumes a peculiar shiny look, which indicates to the workman that the sheet is formed.

The workman then lays the mould upon the plank; takes off the deckle, which he places at his right upon the bridge, and then hands the mould to the coucher. This workman raises it, and places it upon a small, curved, wooden stay in such a position that its inclination will favour the drainage of the water.

The vatman then applies his deckle to the second mould, and makes another sheet. In the meantime, the coucher seizes the mould with his left hand by the short side nearest him, and grasping it again by the upper long side—which, by turning the mould, is on his right—he applies it to the felt, making it describe the quarter of a circle.

When this is completed, the coucher rises, and slides the empty mould along the bridge.

It is after this operation that the vatman pushes forward the second mould deprived of its deckle, which the coucher places upon the curved stay to drain. This workman then takes a felt from the bench on his right, and applies it accurately to the leaf which he has just couched.

The operations just described continue until the post is completed, that is, until the coucher has exhausted all the felts of which it is composed.

The number of the felts in a post is variable; in the older mills it was composed of eight quires, that is to say, 209 felts containing $26 \times 8 = 208$ sheets of paper.

. The two workmen should regulate their work so that the manipulations of each may coincide.

The rapidity of operation, however, depends upon the nature of the pulp and of the wire cloth of the mould.

There are two methods of couching: the French and the Swiss. In the first the coucher applies the mould vertically upon the felt, and then reverses it. In the second, the mould has been rotated 180 degrees, so that it occupies an almost reverse horizontal position; the coucher first leans it on the long side, and then making it describe an arc of 90 degrees, he couches every part of the leaf upon the felt successively.

This last method is more rapid than the first, but requires more skill on the part of the workman to couch the leaves evenly one upon the other.

It should, however, be adopted exclusively in working with pulp not readily drained, as the weight of the pulp causes it to gravitate towards the lower edge, and more or less destroy the homogeneous character given to the sheet by the vatman.

It is important that the coucher should not allow any drops of water to fall upon the pages in raising the mould, which would produce so many spots impossible afterwards to remove.

On beginning a post, it is prudent to place several felts one upon the other, so as to make a softer bed for couching the first sheets. Without this precaution, shreds of pulp remain attached to the mould, and produce breaks in the paper.

Couching by the long side of the mould offers the double advantage of accelerating the work and facilitating the extraction of the pulp from between the laid wires, as each interval will develop the outline of the surface generated by the motion of the coucher.

Another cause of waste proceeding from couching are bubbles, that is to say, blemishes upon the sheet, produced by the interposition of a volume of air between the felt and the sheet, which, in escaping when the next sheet is couched, leaves a depression in the sheet of paper bearing more or less resemblance to a bottle.

New felts are soft, and taking less hold of the surface of the sheet are liable to this kind of accident, which also occurs when the felt is badly stretched.

When the post is completed, the three workmen unite to carry and place it under the press; the central portion being generally more elevated than the edges, the difference in thickness is made up by bits of wood in the shape of an elongated triangular prism.

The workman then inserts a strong piece of oak, from 4 to 7 inches thick, upon which the force of the press is exercised.

It is preferable to employ, instead of the hand-press of the older mills, an hydraulic press, the force-pump of which should be of sufficient diameter to accelerate the operation of pressing. In this case it would be well to place a strong oak plank under the platten of the press, as the elasticity of the wood renders the pressure more uniform.

The requisite degree of pressure attained, the workman brushes off the drops of water that ooze out along the felts, and which would be absorbed by them, and thence by the sheets of paper, when the pressure is taken off.

When the post is carried away, the work of the third hand begins, while the other workmen are filling the vat with a quantity of pulp proportioned to the weight of the previous post.

The layman successively detaches the moist sheets, placing them one upon the other, and throws back the felts upon the bench to the right of the coucher.

The operation of raising the sheets is accomplished in two ways: either upon an inclined or horizontal plane. In the first method, generally employed in France, benches are used with an inclination of about 50 to 60 degrees.

The layman is generally assisted by an apprentice whose duty it is to remove the felt, while the workman lays his sheet upon the preceding one.

This assistant may be dispensed with when the work is done upon an inclined plane, but is necessary when the direction is horizontal. He holds a sort of flat rule, over which the layman throws the sheet as soon as detached from the felt, still holding it with both hands by the two corners of the short side nearest him. The sheet thus supported by the four corners is placed without difficulty upon the preceding one.

Without the aid of the assistant, the sheet, still very moist, would adhere too readily, and it would be difficult to avoid the presence of wrinkles.

When the sheet is adjusted, the assistant withdraws the rule, removes the felt, and the operation is repeated.

If the post has been but slightly pressed, it is difficult to raise the sheets, the pulp has little consistency, and breaks, the corners tear off, and the result is the loss of a great number of sheets. If, on the other hand, the pressure has been too great, the sheets adhere too firmly to the felts and carry away some fibres of wool. It is the duty of the layman, therefore, to see that the pressure is sufficient and neither too light or too heavy.

When all the sheets are detached, the layman takes several felts to cover this first post of paper, and pats it forcibly with both hands to make it into a sort of compact cake, less liable to be torn in handling.

After several days' work, the felts become greasy, and on being detached from the sheets, give rise to a peculiar creak. It is then necessary to wash them with brown soap in order to recover their softness and absorbing qualities, so as to facilitate the operations of couching and separating the sheets.

There also accumulates around the sides of the vat after a certain length of time, a peculiar kind of grease which, if mixed with the pulp, would produce spots upon the paper.

It is advisable to clean the vats at least once a month, or even every fortnight. This cleaning should take place on Saturday, at the close of the day's work. The pulp is used up as far as possible by making a last post of paper somewhat thinner than the others. The pulp is then removed with great care, without touching the substances at the bottom of the vat, which are most impregnated with this grease.

By means of a brush the internal surface of the vat is perfectly cleansed, and, after repeated rinsing, the pulp which had been removed is replaced.

The greasy pulp may serve for making common kinds of paper.

In some mills the posts of paper, after having been subjected to a light pressure, are carried to the drying room. But if it is desired to obtain a superior quality of paper, it is essential to lift and change the sheets.

Lifting consists in detaching one, two, or several sheets together, and forming new posts before hanging them up to dry.

In the change, the sheets are raised one by one, and replaced one above the other in a different order from that which they at first occupied. The workman charged with this work takes alternately a sheet from two posts placed beside him and makes of them a new one, in such a way that the middle sheets are at the outside, and *vice versa*.

The grain of the paper is softened, and it acquires a greater degree of firmness, and assumes that velvety feel which characterises the Dutch papers where this method is universally practised.

After four or five such operations, alternating with pressure applied stronger as the paper gains consistency, the sheets are carried to the drying-room.

These posts of paper being thicker in the middle than at the edges, their irregularity may be compensated by bands of felt inserted when subjecting them to the action of the press.

The paper is hung in spurs of four or five sheets, which are placed upon the ropes or tribbles by means of poles.

Paper, when dried in warm weather, shrinks, and acquires great hardness.

The material when too soft yields to its own weight, expands out, and forms a wrinkle on the back of the sheet—a very objectionable defect. This imperfection is less evident when the paper is delivered folded.

In France, the drying-rooms are placed upon the top of the house. In Holland they are less elevated, and cooler stories are preferred, in order to prolong the drying.

Some manufacturers hang the paper in spurs of seven or eight sheets, in order to economise space in the drying-room. This method is bad in every respect, as the sheets on the outside dry more quickly than the under ones. The air does not circulate, and the drying is not uniform.

The nature of the tribbles very much affects the cleanliness of the paper. Hempen tribbles ought not to be used, as they stain the back of the sheet yellow. Those of aloes and of the fibres of several kinds of cane are advantageously employed: those of horse-hair being the cleanest and best.

With thick tribbles, the air circulates better, and the back of the sheet, being more rounded, is not so apt to crease.

As soon as the paper is dry, which may be known by the peculiar rustling produced among the sheets when they are moved by the hand, the operation of sizing follows.

The duration of the drying depends upon the season, the situation of the drying-room, the diameter of the tribbles, the number of leaves in a spur, the operation of exchange, the thickness of the paper, and the pulp used in making it.

To be kept continuously at work, a well-organised mill ought to be furnished with drying-rooms warmed by hot-air stoves, so as to be able to dry and size the paper at all seasons.

27. Sizing.—To render hand-made paper impermeable, or fitted to receive writing, it is dipped in a warm solution of size or gelatine raised to a temperature of from 68° to 95° Fahr. This constitutes animal sizing.

This size is prepared by dissolving by long boiling the refuse of hides, cartilages, ears, hoofs, tendons, &c., bought from tanners or dealers in them.

This refuse animal matter is dipped in milk of lime, which preserves it from putrefaction, and after drying in the open air is sold under the name of *scrolls, pates, hide pieces, &c.*

Notwithstanding the use of lime, if the desiccation has not been well attended to, this size will undergo a certain kind of fermentation which deprives it of a part of its adhesive qualities. This defect may be recognised by the ammoniacal odour generated while the material is boiling.

Size obtained from young animals, such as lambs, calves, &c., is easily prepared, and of a white colour; whereas that produced by the hides of oxen, cows, &c., is darker, but stronger and tougher, and gives more firmness and resonance to the paper.

To obtain the solution of the gelatine contained in the scrolls, these last are placed in a copper boiler, surrounded by brickwork, and usually heated by a naked fire.

It is well first to wash the scrolls in order to remove the various impurities which they contain.

The duration of the boiling varies according to the nature of the materials. For those obtained from oxen, from twelve to sixteen hours at least are required.

The scrolls should not touch the bottom of the vessel,

as the parts in contact with the metal become too greatly heated, and produce a brown discoloration of the solution. To remedy this objection, in some mills a bed of straw is laid at the bottom of the copper. But though the nature of the evil is changed, the effect remains much the same, as the straw gives out its own yellow colouring matter.

During the first hours of boiling, fatty matters which appear upon the surface must be carefully skimmed off. To facilitate this operation, they are sprinkled with lime, so as to form a calcareous soap.

When it is thought that the solution is sufficiently concentrated, the liquid is run off into a lower vessel, and the copper again filled with water. By again boiling a second weaker solution is obtained, which is generally mixed with the first.

Another method is to let into the boiler a quantity of water proportioned to the volume of the solution drawn off by the stopcock, so that the level of the liquid will remain the same. The boiling is continued, and the fire is removed when it is judged by the amount drawn off that the size-pieces have been deprived of all their soluble principle.

The solution thus obtained is never clear; it contains suspended in it different substances of which it is to be freed. This constitutes the operation of clarifying the size.

A small quantity of powdered lime should be added, and, after being stirred, the liquid is allowed to rest. If the impurities have not completely settled to the bottom, a half a hundredth part of sulphuric acid is added. An insoluble sulphate of lime is formed at once, and its precipitation clarifies the solution.

The liquid is decanted and filtered through several folds of felt, which retain any remaining impurities. This solution contains a mixture of gelatine and chondrine, two substances differing very greatly in their properties. The latter being an obstacle in sizing, it is indispensable to eliminate it. Fortunately, this material may be completely precipitated by a concentrated solution of alum. Sulphate of alumina and sulphate of iron produce the same effect.

The size, when again filtered, is ready for use; if it is too concentrated it may be diluted with warm water. The temperature and the strength of the size both vary according to the nature of the paper and the condition of the atmosphere. Papers manufactured from hard rags require a thin size with a high temperature. The reverse is the case with those made from soft rags. All things equal, it is important that the strength should be greater in summer than in winter.

The papers are dipped in a copper of dimensions varying with the size of the paper.

To keep up an even temperature, the copper should be heated by a small furnace: preferably by the heat of a water-bath or steam, by means of a worm or a copper with double sides.

The workman employed in the operation of sizing, takes a handful containing from 100 to 150 sheets at a time and dips them into the size, separating them like a fan, and manipulates them so that every part of the sheets shall be uniformly saturated. This requires a certain amount of skill.

He then carries the package to a press near the sizing tub.

A series of packages are piled one above the other, and when they have attained a height of about 2 feet, he places a plank upon, and presses them in order to drive out the excess of size, which flows back into the tub placed below.

The degree of pressure should vary according to the nature of the paper; so that, in all cases, the sheets may be readily removed singly and without tearing.

The moist sized paper must then be carried to the drying-room, and hung upon the tribbles in spurs of two, three, or four sheets.

Careful makers exchange the paper when sized before carrying it to the drying-rooms. The drying is benefited by this, and the sheets acquire more strength and firmness.

M. Payen, in his *Chimie Industrielle*, gives the following theory of gelatine sizing:—

"In order that paper may be well sized, it must be properly dried. This should be done gradually and slowly, without, however, being carried so far as to allow the spontaneous decomposition of the gelatine to take place.

"This accident sometimes occurs in summer, especially in damp and stormy weather; the size then liquefies, loses its adhesive qualities, and the operation miscarries. If the drying is too rapid, the size remains disseminated through the entire substance of the paper; but if the process is carried on with proper slowness, the moisture contained in the paper gradually finds its way to the surface, and carries with it the gelatine, which forms a superficial and impermeable coating.

"The drying is moderated by means of ordinary window blinds, the openings of which may be regulated at will. It will be understood that by drying slowly the gelatinous solution is allowed to come to the surface as fast as the evaporation of the water takes place, and that therefore the greater part of the gelatine is concentrated at the surface, and renders the paper impermeable, whereas if dried too quickly this material would remain disseminated throughout its entire substance.

"It is easily ascertained whether the paper is sized on the surface only, by scratching it, and then drawing an ink-mark over the denuded part. The paper sized with gelatine will absorb the liquid, whereas the ink will remain unaffected by machine-made paper sized with resin through the entire thickness."

This explanation is entirely confirmed by the practice of damping the paper when too quickly dried, and appearing to be imperfectly sized. Such paper placed between wet sheets becomes moist, and by again passing through the drying-room, the size disseminated throughout the substance of the sheets comes out on the surface as the water evaporates, and the paper becomes impermeable without any additional gelatine.

This process gives the paper firmness, and some manufacturers were formerly in the habit of always employing it for certain kinds of paper, finding in the excellence of the result a sufficient compensation for the cost of the double manipulation.

The character of the water has a marked effect upon sizing.

It is therefore essential that the water which is to be used in the different manipulations of the pulp should be clarified with much care. If the water is very highly charged with calcareous matters, it is well to precipitate these substances with carbonate of soda, or even alum.

The proportion of these reagents ought, of course, to vary according to the composition of the water.

Certain well waters containing a large quantity of sulphate of lime, absorb 28 grs. of pure dry carbonate of soda, before being enabled to dissolve soap.

River waters require from 150 to 400 grs. of alum to 22 gallons of liquid to effect this precipitate. This fact explains the superiority of the papers produced by some mills situated upon streams with granite beds yielding no calcareous elements.

28. The Sizing-Room.—In a paper-mill, the sizing-room is an apartment set aside for the operation of sizing alone. It contains :—

A furnace furnished with a copper boiler about five feet in diameter by three in depth. It is in this boiler that the size is made, or the gelatine extracted from the substances which contain it.

Also a copper vessel called the sizing tub, in which the operation of sizing is effected. This vessel may be about 3 feet in diameter, and about 2 feet deep. It is mounted upon an iron stand, under which is placed a small stove to keep the size at a proper temperature during the process of sizing.

The tub is placed near the press, so that the size in running off from the saturated paper may return to it, and not be wasted in the passage.

29. Preparation of Gelatine.—Gelatine enters into the composition of the soft and solid parts of animals. In

this condition it is found in the muscular fibre, skin, ligaments, cartilages, and tendons; the membranes contain a large proportion of gelatine, and it constitutes about half the weight of bones.

Gelatine is heavier than water, without taste or smell, colourless, and without reaction upon litmus paper; it is, therefore, neither acid nor alkaline.

Decomposed by fire, gelatine again offers us the same phenomena as these substances, but is easily distinguished from them by certain of its properties.

This material is very soluble in boiling water, but very sparingly so in cold. When two parts and a half are dissolved in a hundred parts of water, the liquid congeals on cooling. The jelly sours in a few days, especially in summer; it then liquefies, and before long takes on all the phenomena of putrid fermentation.

Gelatine or glue is generally prepared for the uses of commerce from parings of hides, parchment and gloves, and from the hoofs and ears of oxen, horses, sheep and calves.

All size is more or less transparent. Some sorts are of a blackish and some of a reddish-brown, while others are of a slightly yellow-white. The most transparent and least coloured are the purest, and these kinds are employed by the paper-maker.

There are several methods of preparing size in our paper-mills. In some of them it is considered sufficient to put the raw materials into a boiler, containing a suitable proportion of boiling water, and to boil them until all the gelatine is extracted. The boiling is kept up from twelve to fifteen hours, for three hundredweight of hide cuttings.

It is evident that the strength of the size is proportional to the quantity of raw material employed.

In other mills a smaller amount of water is, at first, added, and when the decoction is concentrated enough, the liquid is drawn off and replaced by another smaller amount of water. In this manner, three or four solutions are obtained, which are either mixed together or preserved separately.

This method is preferable; because, the pieces of hide being unequally soluble, the first portion of gelatine extracted deteriorates by remaining in the boiler until the rest has been melted. Besides this, the quality of the materials may be better ascertained from the amount of size they have furnished.

Whatever method may be adopted, it is important that the size pieces should not touch the bottom of the boiler, as in that case they would be burnt, and give the size a darker colour.

In the Dutch paper-mills the raw materials are held in a wicker basket, which is let down into the boiler and withdrawn by means of a pulley, when the size is to be removed. This apparatus is very simple, and allows of ascertaining whether there still remains any gelatine undissolved.

With whatever care size is extracted, the decoction is never quite clear; it holds in suspension a great quantity of undissolved gelatinous matter, which would not be precipitated, even after long rest, with the fluidity of the size constantly maintained; but if by any means an abundant precipitate can be obtained, the particles of suspended matter will be carried with it, and the size will then become perfectly clear.

30. Operation of Sizing.—A tub in which the operation of sizing takes place is situated near the press, so that the size which runs off from the handful of sheets just dipped by the workman in the tub, and laid on the bed of the press, may not be lost, but fall back into the same vessel.

The upper surface of the press-stand, which is constructed with a view to great firmness, is raised some thirty inches above the floor. This stand is surmounted by a frame two inches broad all round, and about an inch in thickness. The frame is strongly fixed upon the stand by screw bolts, the heads of which are buried in the frame above, and fastened by a nut under the table. In place of these bolts, strong wood screws may be used, having their conical heads sunk in the frame, which should fit the upper surface of the table so accurately that water may not be able to pass between them. It will be seen that by this arrangement the press-stand presents a regular hollow.

In the left-hand corner of the press-stand, and below the lower surface of the frame, a hole is pierced, slanting upwards, and opening in front of the inner border of the frame, into which is fitted a copper pipe, projecting far enough beyond to pour the excess of fluid continually into the tub. This pipe should be well cemented to the edge of the hole, so that no liquid may be able to run off by any other outlet.

The workman generally stands at the tub, with the press at his right hand, and on his left a bench supporting the sheets, just as they have been carried from the drying-room after the stiffness has been taken out. The tub is then filled with tepid size, and the heat applied under it. A quantity of alum is thrown in, varying according to time and circumstances. The alum is dissolved in hot water, and well stirred to mix it with the gelatine. The alum prevents the size from decomposing, and preserves it for a considerable time. Some manufacturers add white vitriol (sulphate of zinc).

The workman places by his side three or four wooden pallets, by means of which he manages the sheets of paper in the manipulations of sizing. These pallets are pieces of wood, flat upon one face, rounded on the other, and slightly conical at the two extremities. The shape of this implement may be imagined by considering it as formed of a cylinder, three or four inches in diameter by twenty-two in length, terminated at each end by a blunt elongated cone, then cut in two, and by this division forming two pallets.

Standing in front of the tub, the workman takes a handful of paper in his left hand, and supports it from beneath by one of these pallets; he seizes the sheets on the opposite side with his right hand, and takes care to separate them with the fingers of that hand, in order that the size may the more readily penetrate between them; he submerges that entire end of the paper by dipping his hand into the size; he then lifts the bunch with his left hand, and holds it above the tub to drip, which brings the sheets together. The end held in the right hand is allowed to rest upon a second pallet, which generally floats upon the size, while with a third he seizes it above so as to catch it between the two, and lets go the other end, which he held in his left hand. He again separates the leaves with his fingers as before, and plunges his left hand into the size with this end. The bunch of sheets is held suspended for some time to allow the size to run off and the pages to adhere; and the workman, after having raised the lower end with his left hand, carries the lot with both hands to the hollow surface of the press-stand.

This operation is continued until ten or twelve handfuls have been sized. A turn is then given to the press, which causes the size to penetrate the substance of the paper, and the excess of the liquid flows back into the tub by the pipe already mentioned. This operation requires considerable care; for if the paper is pressed too hard, an excessive amount of gelatine is thereby expelled, and the interior of the handfuls of sheets would remain unsized. From experiments made it has been estimated that a quantity of unsized paper weighing eighty-six pounds, absorbs in sizing six pounds of dry gelatine.

31. *Drying after Sizing.*—When the process of sizing is completed, the workman presses the entire mass of paper gently and slowly, not carrying the operation too far, so that the size may have time to become fixed in the substance of the leaves. The paper is only left long enough under the press to ensure a uniform penetration of the size, without allowing it to dry, as in that case there would be danger of the leaves becoming so tightly glued together as to be no longer separable.

When the moment of taking off the pressure has arrived, the paper, still wet, is delivered to women who separate it sheet by sheet, and thus hang it upon the tribbles, in the drying-room, by means of T-shaped lifters, beginning from above. These lifters, with handles varying in length with the height of the tribbles, are employed to avoid the necessity of mounting upon trestles. The entire length of one line is covered with sheets before beginning upon the next.

The Dutch employ a peculiar mode for the preparation of paper after sizing. They introduce the operation known as the exchange, which is done as it leaves the sizing-room, as follows :—

" When the paper has remained long enough under the press, the workman carries it away, in portions of one or two handfuls, and distributes it along the table; he then begins with the nearest lot and exchanges the leaves one by one, lifting them by the corner, so as to form a new pile, which differs from the first only in that the surfaces, which before touched and had been pressed against each other, are made to correspond with the surfaces of other sheets. By thus mixing the sheets in a new distribution each surface is detached from those of the contiguous leaves, to which it adhered, and applied to others, against which it is again pressed. It is of little consequence whether the paper is still warm or not, so long as it is wet. Care must be taken, however, to see that the leaves are not replaced under the press, after the exchange, until the paper is cool ; for if still warm, the size would be fluid and liable to be expelled from the leaves by the action of the press, or to exude unequally upon the surface, thus producing irregularities and destroying the advantage of the exchange.

It will be seen that the process of sizing is one of the most difficult and uncertain operations of paper-making. We are frequently obliged to begin anew, either by the season, the unfavourable condition of the atmosphere, or the temperature of the size itself. If the size is too hot it injures the paper; if too cold it will not penetrate it; if too thick it attaches itself only to the surfaces and runs off when the paper is dry; and lastly, if too thin the sizing will be insufficient. The atmosphere also acts a most important part in the process.

32. **The Process of Finishing.**—Hand-made paper, is the same as that described in the case of paper made by machinery.

On leaving the drying-rooms, the paper is piled up and subjected to repeated pressure in order to remove in part the trace of the fold on the back. A series of packages is made up, between each of which packages is placed a wooden board of the same size as the paper, in order that the action of the press may be more uniform.

Such papers as are not to be rolled require a more heavy pressure. It is also well not to pile them up when they are too dry, as a slight trace of moisture greatly facilitates the pressing.

The nature of the wood from which the boards are made is not a matter of indifference: wood without knots is required; and very dry walnut answers the purpose well.

Sometimes these boards are covered over with glazed board, or the latter is used alone.

Whatever may be the nature of the hard substances placed between the pressings (200 to 500 sheets), it is essential that the pressure should be applied gradually in the beginning, otherwise the sheets crease, or the grain is crushed, and never gives that uniform surface which is characteristic of the products of the best mills.

Before being put up into reams and packed, the pressure of an hydraulic press would much increase the lustre of the paper.

The principal defects which are found in paper made by hand, are knots, water spots, bubbles, folds, torn edges, ragged edges, size stains, holes, and tears.

When the paper is put in reams, it should be perfectly dry, and be laid in a dry and well-ventilated place, otherwise, by the action of moisture, it would soon sour.

33. **Water-marked Paper in general.**—The paper of banknotes, when held to the light, generally presents inscriptions or designs which, in the trade, are termed watermarks.

There are several kinds of watermarks. The most simple are obtained by sewing to the laid or wove wires of the mould fine brass wire twisted according to the outlines of the design, or, still better, a thin copper leaf cut out with a puncher or graver.

The increase of thickness in this place, and the raised wire preventing the pulp from draining, produces a reverse effect upon the paper, the design of which is seen to be lighter than the rest of the sheet.

First, another method is the use of a dark watermark with light letters obtained by depressing the wire cloth itself, in such a manner as to produce a sort of rectangular cavity in which a greater quantity of pulp may be deposited, and then sewing the letters to the bottom of this depression.

Next, shaded watermarks, the richest of all, allowing of obtaining by a sort of moulding process, every variety of relief, of whatever nature.

The principle of depressing the face of wire cloth or the preparation of watermarks, is that every part of the surface should be readily stripped from the paper in order to facilitate the operation of couching.

Moulding the wire cloth, for shaded watermarks, requires the skill both of a moulder and an engraver.

The manufacture of watermarked paper requires especial care. The workman should manage to produce a leaf of exactly uniform thickness throughout its whole extent. This is necessary to secure a clear impression. The coucher should see that this impression is sufficiently distinct, and notify the foreman at once, when a letter or any other part has become detached or unsewed.

Chapter VI.

CHEMICAL ANALYSIS OF MATERIALS EMPLOYED IN PAPER-MAKING.

THE nature of cotton, linen, and hempen rags may be determined by chemical tests, and by microscopic examination when the material is comparatively new.

The fibres of cotton and hemp seen through the microscope, present the appearance of rigid cylindrical tubes, with intercepted intervals like straws, reeds, &c. They are elongated cells, glued together by a material of denser texture, which is susceptible to the action of acids and alkalies.

The diameter of hemp fibres may be estimated to be $\frac{12}{1000}$ to $\frac{13}{1000}$ of an inch. Those of linen are still finer, and have a silkier appearance.

The fibrillæ of cotton, on the contrary, are transparent tubes, flattened in the middle throughout their entire length, presenting the appearance of two parallel rolls, united by a very thin partition. They cannot be better compared than to a T rail, twisted several times upon itself.

The diameters of the fibres vary from $\frac{7}{1000}$ to $\frac{4}{1000}$ of an inch.

When these fibres have been brought into the condition of textures, more or less worn, these microscopic characteristics are insufficient; the straight and rigid conformation of the hemp and flax fibres having disappeared.

It is then necessary to resort to chemical reagents. Linen and cotton fibres may be distinguished in several ways.

1st. A boiling solution of potassa colours the fibres of linen a deep yellow; whereas its action upon cotton is very feeble.

To make use of this test, a piece of the cloth to be tested is placed in a boiling solution of potassa, composed of one part of caustic potassa and one of water. After two or three minutes' immersion, the excess of alkali is expressed by means of several folds of bibulous paper, and the successive fibres of the warp and woof are counted; those of linen being of a deep, and those of cotton of a light yellow, or white.

2nd. Concentrated sulphuric acid quickly attacks cotton fibres, and converts them into gum, while linen fibres remain white and opaque. By washing the gummy matter is removed, and the sulphuric acid neutralised by the addition of a small amount of caustic potassa. The threads of the sample having been counted, the missing ones will represent those of cotton.

In paper-making linen and cotton rags are distinguished by the touch, and the workwomen very soon acquire sufficient skill to make this distinction rapidly.

34. Waters.—The degree of pureness of the waters, used in the different processes of paper-making, has a great influence upon the quality of the papers.

Murky water, containing argillaceous and silicious substances, turns the pulp yellow; that containing saline matters in solution, to a certain extent destroys the brilliancy of the colouring materials, or forms an impediment in sizing with gelatine. If the water contains organic matter in solution, it sometimes prevents the manufacture of superfine papers.

It is therefore important to determine, at different seasons of the year, the composition of the water employed in washing and other manipulation.

We may obtain by filtration the amount of argillaceous and silicious matters held in suspension in the water.

The calcareous salts held in solution are precipitated by means of a few grains of alum or carbonate of soda.

If it is desired to make a complete analysis, the most certain method is to evaporate a quantity of the liquid and examine the residuum after evaporation, for the different proportions of the substances contained.

35. Alkalimetrical Test.—The soda of commerce has no value except in so far as it contains soda in the condition of a carbonate, or of caustic soda. To determine

HENRY WATSON,

HIGH BRIDGE WORKS, NEWCASTLE-UPON-TYNE,

GENERAL MECHANICIAN.

SOLE MANUFACTURER OF HIS IMPROVED KNOTTER PLATES,

Also of improved Revolving and Jogging Strainers in Vats complete,

DOCTOR PLATES, BRASS AND COPPER ROLLS,

HYDRAULIC PRESSES AND PUMPS,

Gun Metal Cocks, Valves, Water and Steam Gauges, Hydraulic Rams, &c.

H. W. begs to intimate that to meet the increasing demand for his improved Strainer Plates he has just completed extensive additions to his Premises and Machinery, and is now in a position to execute all orders promptly, and on the most reasonable terms. The strictest attention will always be given to maintain in the highest degree the quality of the material and workmanship, keeping in view also the capability of re-closing after having been worn.

Rag Chopper, Wheel, 4 ft. diam., 1 ft. wide.

Finishing Calender. These rolls are made of refined Chilled Roll Metals.

BENTLEY & JACKSON,
ENGINEERS, IRONFOUNDERS, AND MACHINISTS,
BURY, NEAR MANCHESTER.

Makers of Paper Making Machinery, Millboard Machines, Paper Cutting Machines, Ripping and Winding Machines, for preparing paper for continuous printing presses.

Hydraulic Pumps and Presses. Steam Engines and Boilers.

ESTIMATES ON APPLICATION.

IMPORTANT TRADE PUBLICATIONS.

Thirteenth Edition. Demy 8vo., 104 pages, Stiff Covers, Price 2s. 6d., or Post free for 30 Stamps,

THE PAPER MILLS DIRECTORY.

Price 2s. 6d., or Post free for Thirty Stamps,

THE PAPER STAINERS' DIRECTORY
OF GREAT BRITAIN.

The Seventh, a New, Corrected, and Enlarged Edition, Foolscap 8vo., price 3s. 6d., or Post free for 30 Stamps,

THE STATIONERS' HANDBOOK,
AND GUIDE TO THE PAPER TRADE.

Sixth Edition, Demy 8vo., 60 pages, Stiff Covers. Price 2s. 6d., or Post free for 30 Stamps,

THE CHEMICAL MANUFACTURER'S DIRECTORY.

Sold by KENT & CO., Paternoster Row.

Price 2s. 6d., or sent Post free for 30 Stamps,

A MAP OF THE PAPER MILLS OF ENGLAND.

Arranged by the Editor of "The Paper Mills Directory."

THE EDITOR, at 24, Bucklersbury, London, E.C.

No. Price

THE ART

OF

PAPER-MAKING:

A Guide to the Theory and Practice

OF THE

MANUFACTURE OF PAPER.

COMPILED BY THE

EDITOR OF THE "PAPER MILLS DIRECTORY."

London:
SOLD BY UNWIN BROTHERS,
OXFORD COURT, CANNON STREET.

B. DONKIN & CO.'S SPHERICAL ROTARY RAG BOILER.

To contain from 20 to 25 cwt. of Rags.

This Boiler being spherical, is twice as strong as a Cylindrical one of the same diameter and thickness. The Plates used are notwithstanding of the usual substance, thus rendering it perfectly safe, durable, and suitable for high pressure steam.

The Spherical shape has also another important advantage, viz. allowing all the rags to fall out by themselves, when the boiler is revolving with its cover off.

Inside the Boiler are Strainers to take off the dirt, and Lifters to agitate the rags during the process of either boiling or washing.

To avoid Cement or run lead Joints, the Gudgeons and the Boiler are turned true in a lathe to fit each other, the joints being simply made with red lead.

B. DONKIN & CO.,
ENGINEERS, MILLWRIGHTS, MACHINISTS, AND IRON FOUNDERS,
Blue Anchor Road, Bermondsey, London, S.E.

Manufacturers of Steam Engines, Turbines, Water Wheels, Flour Mills, &c., Paper Making Machines, and everything connected with Paper Mills, Rag Engines, Rag Cutters, Cutting Machines, Glazing and Hydraulic Presses, Pumps, Hoists, &c., Steel Bars and Plates, Wires, Felts, and Deckle Straps. Machines for Printing in two Colours. Gas and Water Valves.

the quantity of these materials present, resort may be had to tests with the alkalimeter.

The principle upon which experiment is based is as follows : Given a dilute solution of free alkali, of carbonate or sulphate of potassa or soda, of chloride of potassium or sodium, &c. Pure sulphuric acid, diluted with water, is added to the mixture ; this acid acts only on the free alkali, or on the carbonate, and, as long as the acid is not present in sufficient quantity to produce a neutral sulphate, the liquid manifests an alkaline reaction ; but when the base is saturated, the solution becomes neutral to coloured tests, and when this point is exceeded never so little, the liquid will redden litmus paper. We are thus enabled to determine the exact moment of saturation.

Experiment shows that, if the substance analysed is pure potassa, it would take 77 grs. of sulphuric acid to neutralise 74 grs. of potassa.

Supposing that we are operating on the potash of commerce, containing water, carbonic acid, chloride of potassium, and sulphate of potassa. If instead of 77 grs. 35 are sufficient, this potash contains 50 per cent. of foreign and worthless matters.

To make this experiment, 741 grs. of the potash to be tested are weighed off and dissolved in such a quantity of water that the solution shall occupy 30 cubic inches. Take $3\frac{1}{2}$ cubic inches, and this volume will contain in consequence 74 grs. of the potash to be tested.

This alkaline solution is then poured into the vessel in which it is to be neutralised.

The sulphuric acid is prepared by dissolving 154 grs.

of sulphuric acid in a sufficient quantity of water to complete the volume of 1·76 pt.

If we now take a burette, graduated 100 divisions will correspond to 77·20 grs. of pure acid, and from this it results that, if it takes 100 divisions to complete the saturation, the potash under consideration would be pure. For 60 divisions the proportion of alkali would be 60 per 100.

The number of divisions, or degrees, on the alkali-metrical burette expresses, therefore, the proportion by weight of the alkali contained in the material to be examined.

36. **Examination of Limes.**—The principal substances to be met with in limes are magnesia, silica, alumina, and traces of the oxides of iron.

The limes to be preferred for use in Paper-making are the rich white kinds, which contain at least 90 per cent. of pure lime.

To make an analysis of any given lime, take 154 grs. of the material, and heat them in a porcelain capsule, at a temperature of 212° to 392° Fahr. The difference of weight now represents the proportion of moisture contained in the specimen. Raising it to a red heat in the same capsule, or better still in a platinum crucible, we determine by again weighing the amount of carbonic acid which might have been left in combination with the lime through imperfect calcination.

Then take 30·88 grs. of this material and treat them with dilute hydrochloric acid, gently heating to complete the solution of all soluble principles.

By pouring ammonia into the solution, the alumina,

which might have been dissolved by the hydrochloric acid, will be precipitated.

The alumina and silica are separated from the solution by filtering, and then weighed after desiccation and incineration of the filtering paper, the ash of the paper being subtracted from the weight.

In the filtered liquid the lime is precipitated either by oxalate of ammonia or by sulphuric acid. In the first case by calcining we obtain caustic or quick-lime, which is weighed. In the second, the lime is treated in the condition of a sulphate, and after precipitation of the magnesia by the phosphate of ammonia and magnesia, is calcined, and the phosphate of magnesia weighed.

By means of chemical equivalents the proportions of caustic lime and magnesia contained in the material are determined.

As the caustic portion is alone fitted for the uses of alkaline boiling, it should alone determine the market value of lime; the other substances being, if not injurious, entirely inert.

For manufacturing white paper, perfectly white limes should be used. This quality, indeed, is an indication of their pureness, as poor limes are of a more or less greyish colour, owing to the large proportion of clay they contain.

36.* **Chlorometric Tests.**—We owe the chlorometric process generally followed to Gay-Lussac; it is based upon the oxidising properties of chlorine salts in the presence of water; the water in decomposing yielding its oxygen to the oxidisable body and the chlorine uniting with the hydrogen to form hydrochloric acid; and further, upon the instantaneous discoloration of a solution of indigo by a slight excess of chlorine.

Arsenious acid taken as the oxidisable body becomes converted into arsenic acid.

The test liquid composed of arsenious acid, which is bought ready prepared at manufacturers of chemicals, is such as to necessitate a volume of a solution of chlorine equal to its own, in order to convert all the arsenious acid into arsenic acid.

This liquid contains 67 grs. troy of arsenious acid to 1·76 pints of the solution. It is prepared by dissolving that quantity of arsenious acid in pure hydrochloric acid, diluted with its own volume of water, and afterwards adding water enough to complete the litre.

To perform the experiment, take 154·42 grs. of the chloride to be examined (chloride of lime, for instance), and wash them well with water several times in a mortar; throw all the water used in this washing into a flask and add more water to make up the 1·76 pt.

Pour 0·61 cub. in. of the arsenious solution into a vessel, slightly coloured by a solution of indigo. The burette is then filled with the chlorinated solution. The scale is graduated in such a manner that a hundred divisions correspond with ten cubic centimetres and the test liquid poured in drop by drop until the colour disappears.

In order better to determine the instant at which the discoloration occurs, the vessel is placed upon a sheet of paper.

If 100 divisions are used the degree would be 100.
,, 200   ,,   ,,   ,,   ,,   50.
,, 150   ,,   ,,   ,,   ,,   66·6.

100 degrees indicate that 2·679 lbs. troy of the chloride contains 6102·70 cub. in. of chlorine gas.

The chlorides of commerce contain about 5049·67 grs. of chloride to 2·679 lbs.

It is well, as in the alkalimetric tests, to perform two or three successive experiments, in order to ensure the accuracy of the first.

When once accustomed to testing in this way, a person can make several analyses of different chlorides in an hour.

The following, according to M. Payen, is the composition of chloride of lime.

1st. Pulverised, at 100°, that is, containing 100 volumes of chlorine.

Composition by equivalents :—

$$\begin{array}{rcl} 2 \text{ Cl.} & = & 886\cdot4 \\ 4 \text{ CaO} & = & 1400\cdot0 \\ 4 \text{ HO} & = & 450\cdot0 \\ \hline & & 2736\cdot4 \end{array}$$

Or by hundredths :—

Chlorine . . . . . 32·39
Lime . . . . . . 51·16
Water . . . . . . 16·45
───────
100·00

2nd. Composition of chloride of lime in aqueous solution.

$$\begin{array}{rcl} 2 \text{ Cl.} & = & 886\cdot4 \\ 2 \text{ CaO} & = & 700\cdot0 \\ 4 \text{ HO} & = & 450\cdot0 \\ \hline & & 2036\cdot4 \end{array}$$

Or by hundredths :—

Chlorine . . . . . 43·52
Lime . . . . . . 34·37
Water . . . . . . 22·11
———
100·00

These figures clearly establish the fact, that the same quantity of lime will absorb twice the amount of chlorine, when the reaction takes place in the presence of water. There is then advantage in preparing the liquid chloride, when it is not necessary to transport it. The powdered chloride may be considered as retaining the chlorine by a double equivalent of lime, whereas in the liquid chloride it is the excess of water which, in effecting the solution, preserves the stability of the compound.

36.** Examination of Manganese.—The market value of this substance, as applied to Paper-making, depends upon the amount of chlorine it is capable of liberating in the preparation of chlorine gas, or the decolorising chlorides. The weight of chlorine disengaged is proportional to the quantity of pure binoxide of manganese contained in the ore.

To ascertain the value of this substance, it is sufficient to act with it upon an excess of hydrochloric acid, and then determine the quantity of chlorine disengaged and collected in an alkaline solution.

Precise analyses have allowed us to determine that 61·44 grs. troy of pure binoxide of manganese disengage 6102·70 cub. in. of dry chlorine at 278° 0; and under a pressure of 20 inches of the barometer, or one atmosphere.

It is evident from this that the hundredths of a litre of chlorine will represent the hundredths of available manganese contained in the specimen of ore.

To perform the experiment, therefore, weigh 61·44 grs. of the manganese, reduced to powder, place it in a flask, and gradually add 1·52 cub. inches of hydrochloric acid through an S-shaped tube. The flask is supplied with another tube, through which the gas is conveyed into the alkaline solution.

The mixture is heated gently, and the operation continued, until the vapour of water has driven off the last traces of chlorine.

The resulting solution of chloride is poured into a flask, bearing an index marking the capacity of 1·76 pt. This measure is completed by adding the water used in rinsing the flask, from which the liquid was taken.

The test liquid being thus obtained, the experiment is concluded by employing the ordinary chlorometric test.

The value of manganese also depends upon the proportion of hydrochloric acid which it employs in liberating the chlorine. The binoxide of manganese (MnO_2) requires two parts of hydrochloric acid to disengage one part of chlorine; the sesquioxide (Mn_2O_3) three parts of acid for one of chlorine; and finally the protoxide (MnO) takes one part of acid, with which it simply forms a chloride, without any evolution of gas whatever.

These different reactions are expressed by the following chemical formulæ:—

$$MnO_2 + 2HCl = MnCl + 2HO + Cl$$
$$MnO + HCl = MnCl + HO$$
$$Mn_2O_3 + 3HCl = 2MnCl + 3HO + Cl.$$

The carbonates of lime and baryta, and the oxide of

iron, which are found in manganese ores, also combine with the hydrochloric acid, causing a proportional loss in the amount of gas evolved. The result is the liberation of carbonic acid, which, in the manufacture of chloride of lime, forms a carbonate which the chlorine will not decompose. We thus have a loss in acid, in lime, and in the degree of the chloride. M. Payen, in his *Chimie Industrielle*, gives the following process for determining the amount of acid employed.

To dissolve 61·46 grs. troy of the binoxide of manganese, and produce 61·02 cub. in. of chlorine, a quantity of hydrochloric acid equivalent to 135·10 grs. of concentrated sulphuric acid is required; half of the hydrochloric acid, or 88°, forming chloride of manganese, and the other half giving the 100° of chlorine. One-hundredth of the acid lost in the operation is replaced.

To determine the amount of hydrochloric acid required by any specimen of manganese, treat 61·46 grs. with 1·52 cub. in. of acid, representing 250 acidimetric degrees (equivalent to 193·02 grs. of concentrated sulphuric acid) and collect the chlorine. Admitting that 100 chlorometric require 173 acidimetric degrees, it is ascertained by neutralising with carbonate of soda (until the carbonate of manganese ceases to be dissolved) how much of the free acid remains. Now, by adding these two quantities we determine how much is needed to complete the 200 degrees of acid employed, and the deficiency will represent the amount of loss occasioned by the specimen of manganese.

37. Antichlorine.—We saw, while speaking of the washing of the pulp, that in order to remove the last traces of

chlorine retained with much energy by a sort of special capillary attraction, it was advantageous to make use of alkaline sulphites, or hyposulphites, which, by combining with the chlorine, annihilate its destructive effects.

Antichlorine is used in England upon a large scale. It has been calculated that the annual consumption amounts to 196·85 to 321·45 tons.

To ascertain whether the pulp contains free chlorine, different reagents may be used, all, however, based upon the action of iodine upon starch when expelled by chlorine from its ioduretted compounds.

The test liquid may be prepared in several manners, of which we will here give two.

1st. Carry the following mixture to the boiling point :—
    1 part of iodide of potassium.
    2 parts of starch.
    3 parts of water.

This liquid, preserved in a glass-stoppered vial, will colour the pulp blue, if it contains chlorine.

2nd. Boil for about three-quarters of an hour—
    Starch . . . . 5 parts.
    Fused chloride of zinc . . 20 ,,
    Water . . . . 1000 ,,

When the liquid is cool, add—
    Iodide of zinc . . . . . 2 parts.

To ascertain the presence of chlorine in the pulp, a bolus is made of the pulp by squeezing out the greater part of the water. The presence of chlorine is then manifested by the violet-blue colour of the iodide of starch.

There is a great advantage in using the hyposulphite instead of the sulphite ; indeed,

2·679 lbs. troy of sulphite of soda absorb 4339 grs. of chlorine.

2·679 lbs. of hyposulphite of soda absorb 17·666 grs. of chlorine, or about 12·6903 cub. ft.

It follows from this that to counteract 11·1244 cub. ft. or 2·679 lbs. of chlorine, it would take—

9·464 lbs. of the sulphite of soda.

2·341 lbs. of the hyposulphite of soda.

Since the cost is equally in favour of the hyposulphite, this compound should be considered as the practical antichlorine for all industrial purposes.

38. Alums.—Different kinds of alum are met with in commerce.

1st. Alum with a potassa base, of which the formula is—

$$KO,SO_3 + Al_2O_3 3SO_3 + 24HO,$$

presents on analysis the following composition :—

Potassa	10·82
Alumina	9·94
Sulphuric acid	33·77
Water	45·47
	100·00

2nd. Alum with an ammonia base, of which the formula is—

$$(NH_3HO)SO_3 + Al_2O_3 3SO_3 + 24HO,$$

gives the following analysis :—

Ammonia	3·89
Alumina	11·90
Sulphuric acid	36 00
Water	48·21
	100·00

Alums contain iron in variable proportions, and the presence of this metal is injurious to the colour of pulps of delicate and pure shades. For such purposes it is well to use the purified material, known as refined alum, and proof against the tests of prussiate of potassa, which reveals the feeblest traces of iron. It is sufficient to pour a few drops of ferrocyanide of potassium (yellow prussiate) upon some crushed alum, to obtain a blue tinge, if the alum contains iron.

This method is also employed for purifying alum.

Into a solution of alum enough of the prussiate is poured to precipitate the whole of the iron; after allowing it to rest, the liquid is decanted by means of a syphon, and may be used at once or reduced to the form of crystals.

The precipitate of Prussian blue may be used for colouring pulp.

Instead of alum, sulphate of alumina is also used, which is generally obtained by roasting together aluminous schist and iron pyrites, and is apt to contain considerable quantities of that metal; so that it is necessary to purify this substance with ferrocyanide of potassium, when it is to be employed in making fine-coloured papers.

Although alum is more costly than sulphate of alumina, it is preferable because less variable in composition. The sulphate of alumina, which sometimes contains an excess of acid varying from two to six per cent. of the weight of the sulphate, gives rise, by this irregularity of composition, to considerable practical difficulty.

39. Kaolin.—Kaolin is a basic silicate of alumina,

produced by the decomposition of feldspathic rocks. In a crude condition it contains sand of various fineness, of which it must be freed before using in Papermaking.

This purification is effected by successive washings, which carry off the finest particles. This clay, although smooth to the touch, when dry adheres roughly to the tongue.

We give the composition of washed kaolin :—

Water	12·82
Silica	48·37
Alumina	34·95
Oxide of iron	1·26
Potassium and soda	2·60
	100·00

To ascertain the pureness of kaolin, it is only necessary to wash it and then determine the quantity of quartz granules contained in the specimen.

It is always prudent not to employ kaolin in the preparation of size until it has been passed through the meshes of a fine sieve.

40. Starch.—Starch is a mealy substance, generally extracted from potatoes. The chemical formula for that of commerce is $C_{12}H_{10}O_{10} + 4HO$.

Starch contains variable proportions of water. What is called dry starch, or that containing four equivalents, has 18 per cent. of its weight made up of water. Placed in a very moist atmosphere it may absorb 10 equivalents, or contain 35 per cent., while other starch which has only been drained contains 45 per cent. of its own weight of water.

It is, therefore, important to make sure of the hygrometric condition of the starch by desiccating it at a temperature of 68° to 86° Fahr. in a dry atmosphere. To increase the weight of starch it is mixed with various matters, such as chalk, plaster, sulphate of baryta, &c. The adulteration is, however, easily recognized by incineration. Starch burns up entirely, leaving no residuum, whereas the incombustible mineral matters are found at the bottom of the porcelain capsule in which the experiment is performed.

41. — Colouring Matters. — Colouring materials can scarcely be otherwise tested than by directly staining a certain fixed weight of pulp. A well-supplied paper-mill ought to have a small cylinder engine, holding about 11 lbs. avoir. of dry pulp. The quality of the colouring matter may then be determined by gradually increasing the proportion and making several sheets of paper on a small mould. The results obtained in this manner are then compared. This method of experiment renders great service when a new kind of paper is to be made, and we have no very definite idea of the amount of colouring matter necessary to give the required shade.

M. Liverkus, a manufacturer of ultramarine in Germany, recommends the following very practical method, applicable, however, to this substance alone.

After having poured 46·32 grs. troy of the ultramarine to be tried into a vessel containing 4,168 grs. of a concentrated solution of alum, the material is agitated so as to cause all the colouring matters to become suspended in the liquid.

After resting for some time, say half an hour, the mixture is again stirred.

The colouring power of the ultramarine will be directly proportional to the length of time required for the solution to be discoloured.

With some blues the discoloration takes place at the end of three to six hours. In others, on the contrary, no perceptible change in the colour can be observed until the second or third day.

The aluminous solution is prepared by dissolving 1,135 grs. of alum in 2·679 lbs. troy of water.

Ochres have a commercial value proportional to the care with which they have been washed. It is an easy matter to ascertain the amount of gravelly material they contain, and the same holds good in regard to clays, which are used in considerable quantities for manufacturing common papers and boards. It is, however, important not to pour these matters into the engine, or the vat, until they have been filtered through a long-napped felt, which will retain the coarser particles. For the finer colouring matters, fine felts or flannels are used.

42. Fuel.—The analysis of combustibles has for its object to determine the proportion of ash contained in them, and their calorific power; that is to say, the number of degrees of heat which may be generated by a given quantity of fuel.

The weight of the ash may be obtained by calcining 30·88 grs. troy of the combustible to be examined in a platinum crucible, and stirring the material several times, in order entirely to consume the carbon. The analysis may be considered perfect, when no difference can be observed between two consecutive weighings.

The ash containing much carbonate of lime, which is transformed by calcination into caustic lime, a few drops of a solution of carbonate of ammonia should be added and the whole carried to a red heat.

The results of this experiment are generally considered sufficient, as they allow us to form an estimate of the commercial value of the fuel. It is, however, also of the greatest importance to have correct notions of the number of calorific degrees which can be obtained.

The examination with litharge is the simplest method, and is based upon the following principle :—

The amounts of heat emitted by different combustibles are to each other as the respective amounts of oxygen absorbed by these combustibles in burning.

It will be sufficient then to compare the amounts of oxygen required for the combustion of various specimens of fuel with that already determined for pure carbon.

Take 15·44 grs. troy of the pulverised fuel and mix it with 308·84 grs. of litharge. This mixture is placed in a crucible, and covered over with 463·23 or 617·69 grs. of pure litharge. The capacity of the crucible ought to be such that the volume of the material should not occupy more than one-half in order to allow for the swelling which takes place As soon as the fusion is completed the substance is allowed to cool after having been heated briskly for ten minutes. The crucible is then broken, and the lump of lead which is found in it weighed. The calorific power of the fuel is proportional to the weight of the lead. This hypothesis, which is not mathematically exact, gives a sufficient approximation, however, for all practical purposes.

It is indispensable to use litharge free from minium (red oxide of lead).

Pure carbon produces with litharge thirty-four times its own weight of lead, and according to experiments made by M. Depretz each part of lead is equivalent to 230 degrees of heat.

Pit coal is the fuel generally employed in France. The calorific power of this material is somewhat variable according to its source and the proportion of ash it contains.

We give, for the sake of information, the following results of analyses:—

Source.	Ash per cent.	Calorific power.
Hartley (England)	1·50	6781
Anthracite (Wales)	1·60	7406
Mons (Belgium)	2·16	7297

43. Examination of Papers.—Gelatine-sized papers are usually recognisable by their odour; this characteristic is not, however, sufficient for some of those sized in this manner but made by machinery.

By incineration they are found to burn badly, leaving a very black carbonaceous deposit; whereas paper sized with resin in the pulp yields rather a greyish residuum. If the results obtained are uncertain, we must then have recourse to an elementary analysis. Gelatine will be indicated by the amount of nitrogen collected.

As resinous-sized papers always contain starch in

greater or less quantities, a blue colour is obtained when they are subjected to the vapour of iodine.

Generally it is thought sufficient to determine the amount of mineral matters which the paper contains. All that is required, then, is to incinerate 154·42 grs. troy in a porcelain capsule, and to weigh the remainder after calcination.

Paper made exclusively of rags yields a residuum of a third to a half or one per cent. at most, proceeding from the ash of the textile fibres, and the small amount of mineral matters introduced by the waters.

By deducting a hundredth of the weight of the residuum, the weight of the additional matters is obtained.

The substances generally employed are—

Kaolin.
Chalk or carbonate of lime.
Sulphate of lime.
Sulphate of baryta.
Clays.
Ochres.

The examination of these different substances necessitates a complete analysis, which is more in the line of the chemist than the manufacturer of paper.

We give the course to be pursued, based upon experiments made to determine the nature of substances composing the size used in pulps, or resinous sizing.

Chemical Examination of Paper Sized in the Pulp.

" Subjected to the following tests: Boiled with pure water, it yielded a liquid which restored the blue colour to red litmus paper, thus revealing the presence of an alkali. An infusion of nutgalls scarcely affected its transparency, so that it did not contain gelatine; but iodine produced a very intense blue colour, indicating that starch formed a part of the composition.

"186·28 grs. troy of the same paper were boiled for about a quarter of an hour in water acidulated with sulphuric acid. The liquid was expressed through a piece of fine linen and the pulp well washed with boiling water. When dried it weighed only 172·31 grs. The acidulated liquid was united with the water used in washing the pulp, and saturated with carbonate of lime. After being filtered it was partially evaporated, in order to remove the greater part of the resulting sulphate of lime; when evaporated almost to dryness, a yellow residuum was obtained, having a gummy appearance, and weighing 10·33 grs. This substance, when heated in a platinum crucible, became swollen, emitted an odour of burnt bread, and gave an ash containing sulphate of lime with the sulphate of a fixed alkali, not determined.

"The solution in water of this apparently gummy substance was only feebly precipitated by an infusion of galls; but assumed a beautiful dark violet colour when treated with iodine. This material was, therefore, only slightly modified starch.

"The 172·31 grs. of paper which had resisted the action of the boiling water acidulated with sulphuric acid, were treated with a weak solution of potassa. The expressed liquid, while boiling, was of a transparent yellowish colour, but became somewhat turbid on cooling, and gave a lather like soapsuds.

"A small quantity of dilute sulphuric acid was poured into this liquid to neutralise the potassa, when it became very milky and deposited a flaky matter, which did not redissolve by heat. This weighed about 3·08 grs. after evaporation in a capsule smeared with grease. The capsule, as well as the flaky matter, was washed with alcohol, which assumed a brownish colour, and became charged with fatty matter.

"The residuum, insoluble in alcohol, was composed to a great degree of starch which had escaped the action of the boiling acidulated water. The liquid, separated from the 3·08 grs. of flaky matter by sulphuric acid, also contained starch; for when evaporated in order to crystallise the greater part of the sulphate of potassa, it yielded a yellowish mother-water, which gave a deep blue colour with iodine, and a brownish sediment was formed still containing starch. When distilled in a test-tube it gave an alkaline liquid which turned red litmus paper blue.

"This appears to be due to the gluten contained in the flour of cereals used to size the paper under consideration.

"To return to the brownish alcoholic solution resulting from washing the flaky matter. After evaporation there remained 1·54 grs. of a fatty and somewhat pitchy substance, of a yellowish-brown colour, and having about the consistency of lard. Its combination with potassa was very highly coloured and had a bitter taste, which seemed to indicate the presence of a resin. To ascertain whether the suspicion was correct, it was treated with water and a small quantity of magnesia to neutralise the fatty acids, and then subjected the residue to the action of boiling alcohol, which, on evaporation, left behind a slight coating of varnish, easily recognised as a resin.

"77·20 grs. of paper yielded when burnt, 0·92 grs. of a very ferruginous ash, which also contained a noticeable quantity of manganese; for when melted with soda in the flame of a blowpipe, it gave a beautiful blue glass. This ash did not effervesce with acids. When heated to redness with sulphuric acid, and the residuum treated with water, it had very little taste at the time of mixture, but at the end of twenty-four hours the liquid had acquired a very distinct astringent taste, and with ammonia produced a gelatinous precipitate of alumina; from which it follows that alum had entered into the composition of the paper pulp."

A great many machine-made papers become yellow, or present spots of a yellow rust colour. These spots are due to the action of the chlorine contained in the pulp upon the iron of the drying cylinders. Indeed the felts soon become spotted with the same colour at different points, and sometimes over their whole surface. It may be shown that these spots are really due to an oxide of iron, by employing the sulphocyanide of potassium, which produces a red colour, growing deeper as the proportion of the peroxide of iron is increased.

By digesting the leaves of paper to be tested in dilute hydrochloric acid, the paper becomes white, and the presence of peroxide of iron is detected in the solution.

MM. Fordas and Gelis have published a paper upon this subject in the *Journal de Librarie*, from which we extract what follows:—

"On leaving the rag engines the wet pulp is conveyed into a wooden vat, and thence to the various apparatus which constitute the paper-machine, and whose objects are the formation and drying of the paper, and its division into leaves.

"Perfectly washed pulp would be in no way changed during these different operations; for the vapours generated being only those of water, could not, by acting on the different materials of which the apparatus is composed, give rise to any soluble compound capable of becoming incorporated with the leaf during its manufacture.

"But instead of supposing such a case, which is never practically attained, an incompletely washed pulp is used; a large part of the chlorine it contains will, it is true, be carried off by the excess of liquid during the first stages of the operation; but there will always remain a portion, which will be liberated with the vapour of water at the time of drying, attack the cast-iron rollers, dissolve their surface, and form with them a minimum chloride of iron, with which the felts supporting the leaves will become impregnated, and which from them will be introduced into the substance of the pulp itself.

"This impregnation of the felts with a salt of iron cannot be denied. These felts are always spotted with rust, and the yellow colour begins to be perceptible during the first days they are in use. The rust actually becomes combined with the tissue, and is the result of a maximum basic salt of iron, produced by the action of the air upon the minimum salt we have before mentioned.

"This salt cannot possibly affect the paper; it is insoluble and combined; but it is the free and soluble part which exists upon the surface of the cylinders, or in the substance of the felts.

"We will admit that the ferruginous compound enters the paper in a soluble condition, and at the minimum of oxidation, because this fact seems to us to be proved by the absence of colour in the paper at the time of manufacture.

"The complete state of dryness of the paper maintains the salt of iron for some time at a minimum, and consequently in a colourless condition; but very soon the oxygen of the air, assisted by atmospheric moisture, reacts upon this compound, and, by bringing it to a maximum degree of oxidation, colours it.

"This simple reaction perfectly explains the yellow, and often nankin colour which these ferruginous papers assume. It also explains an observation made to us by a printer, namely, that this colour is frequently produced when the paper is wetted for printing, or accidentally.

"As for the round spots, which have more particularly engaged the attention of manufacturers, they may be quite as easily explained. We attribute them to a phenomenon of crystallisation.

"They are formed through the tendency, possessed o a certain degree by the molecules of all bodies, to arrange themselves in groups when they are disseminated throughout a permeable medium. If the spots of which we are speaking are closely observed, it will be remarked that each one of them encloses near its middle an asperity, or hard body, which seems to have served as a centre of attraction.

"As a result of the condition of alternate dryness and moisture to which the porous and hygrometric material of paper is exposed, an insensible displacement occurs. The molecules of the salt of iron arrange themselves around the most compact parts of the paper pulp, as a salt in concentrated solution is deposited around the glass rods or strings suspended in it; or, to make use of a comparison, which seems to us more exact, these ferruginous molecules group themselves in the sheet of paper, as we find them doing in wet soils to form those globules of oxide of iron, or radiated pyrites, so abundant in all localities where ochrous earths are buried in organic deposits.

"Frequently when the displacement we have just explained does not occur until after printing, it is the printed letters themselves which become the centres of attraction, so that the iron becomes fixed in preference upon the printed portion of the paper, which it deeply stains, while the margins appear relatively colourless."

44. The Laboratory.—*Instruments and Apparatus.*

Microscope.
Hand lens.
Scales with stand, weighing as low as 0·154 grain troy.
Blowpipe.
Stationary furnace.
Hand furnace.
Porcelain capsules of different diameters.
Platinum crucible and accessories.
Porcelain mortar.
Spirit lamp.
Instrument for ascertaining the specific gravity of salts.
Thermometers, graduated upon glass.

Earthen crucibles.
Test tubes, with stand.
Flasks.
Glass tubes, straight.
„   „   curved.
„   „   S-shaped, with funnel end.
Glass funnels.
Earthen funnels.
India-rubber pipe for joints.
Wooden tongs and holders.
Alkalimetric apparatus.
Chlorometric   „
Acidimetric   „
Filtering paper.

Reagents.

Blue test paper.
Red   „   „
Hydrochloric acid.
Sulphuric   „
Nitric   „
Acetic   „
Ammonia.
Iodine.
Potassa.
Soda.
Lime.
Oxalate of ammonia.
Carbonate of soda.
Carbonate of ammonia.
Phosphate of soda.
Alum.

Protochloride of tin.
Acetate of lead.
Ferrocyanide of potassium.
Sulphocyanide of potassium.
Sulphate of iron.
Sulphate of copper.
Litharge.
Tannin.
Starch.
Alcohol.
Ether.
Arsenious acid test liquid.
Sulphuric acid ,, ,,
Hydrotimetric ,, ,,
Borax.
Salts of phosphorus.
Chromate of potassa.
Iodide of potassium.

Chapter VII.

WASHING-WATER, WATER POWER, STEAM BOILERS, &c.

45. Washing-Water.—The water which is used for the preparation of the pulp, especially for washing purposes, is called *wash-water*, in distinction to that which only drives and produce power.

To perceive its great importance we need only consider the quantity which is necessary to wash from 400 to 500 pounds of rags in an engine. If the engine is, for example, 15 feet long, 7½ feet wide, and filled about 2 feet high, it will hold, taking the rounds ends, backfall, &c., into consideration, about 160 cubic feet or 1,200 gallons, or about 10,000 pounds of water—that is, about twenty times the weight of rags. If enough water is used during the operation of washing to fill the engine five times, the 500 pounds of rags will be brought in contact with 6,000 gallons, or one hundred times their own weight of water. This quantity may yet be considerably increased, perhaps doubled, if we add the water which is used in boiling, bleaching, and on the paper-machine.

Every pound of rags is therefore liable to be soiled during its transformation into white paper by the impurities contained in 100 to 200 pounds, or in 12 to 24 gallons, of water.

These impurities are of two kinds: those which are only suspended, floating, or *mechanically* mixed, and others fully dissolved. In most cases the latter cannot be seen, and, to make a distinction, may be called *chemical* impurities.

46. **Mechanical Impurities.**—If a stream flows through soil composed of clay or other soft material, it will be clear as long as nothing disturbs its quiet and even flow; but as soon as rain falls, some of the infinitely small and light components of the earth are bodily carried along by it, and the stream becomes turbid and coloured. All these mechanical impurities are visible separate matters, and which will settle on the bottom if allowed time to do so.

Mills which use surface streams as wash-water should therefore be supplied with settling ponds, of as large dimensions as the locality will permit. The water is admitted into these reservoirs when it is clear and shut off when muddy; they should therefore be large enough to hold many days' supply.

If a large settling pond cannot be had, the mechanical impurities must be separated by filtration. The cheapest and most permanent materials for filters are gravel and sand. They are hard, principally quartz, and their round form prevents them from forming a mass so compact that water cannot pass through, while they present at the same time a very large surface for the deposition of impurities.

The filters may be built of brick or stone and cement, or simply of earth; their bottoms, according to Planche, are to be covered with coarse gravel or stone, ordinary gravel succeeding, and sand forming the upper layer. The water usually enters on the top and leaves at the bottom. In the course of time such an amount of dirt will settle on the stones that water can no longer be purified by passing through them, and then they have to be thoroughly washed.

To suffer no delay, the mill should be supplied with two such filters, or with one divided by a partition, which may be opened or closed at will, so that one part can be cleaned while the other is yet in operation.

A useful kind of filter is a square brick filter, divided by cross walls into four equal compartments, which are filled several feet deep with fine gravel, and connected by short pieces of large iron pipes in such a manner, that the water passes constantly through three of them in succession, while one can be cut off and washed out. A workman enters for this purpose, moves the gravel to one side, thus making an empty space, on which he gradually washes the gravel by mixing it with plenty of water. The dirty water escapes through a valve with which each chamber is provided.

The wash-water frequently passes through additional strainers before entering the reservoir in the upper part of the mill. One, which is frequently seen, consists of two wire-cloth covered frames, with woollen rags, slightly cut in the engine, filled in between them.

This double frame is fastened horizontally in a tub or vat, the water passing through it from the lower side, so that the impurities cannot lodge upon it, but will fall to the bottom and leave the wire unobstructed.

If the water enters the engine from the top, a flannel bag may be tied around the outlet of the pipe, and if frequently washed out it will be of some assistance.

47. Chemical Impurities.—The chemical impurities, mostly invisible, are as numerous as the materials over which the streams pass in their courses. Some metals, but especially iron, some of the alkalies, principally lime, and extracts of decaying vegetable matter from the drainage of cultivated fields, are the substances which are mostly found dissolved in water.

Carbonates of lime and magnesia are very slightly soluble in pure water, but dissolve freely in water which contains carbonic acid, and as the latter is always to some extent absorbed from the air, the water is capable of holding some carbonate of lime or magnesia in solution.

Sulphate of lime or gypsum is another form under which this alkali is found in water.

Common salt or chloride of sodium appears occasionally.

Water which contains much lime or magnesia is called *hard*, and every housekeeper knows that it will not answer for washing purposes, as it does not dissolve soap until the lime or magnesia has been precipitated.

If the carbonic acid, which enables the water to hold the carbonates of lime and magnesia in solution, is driven out by boiling, or absorbed by caustic lime or soda which may have been added for this purpose, the carbonates will assume a solid form, settle to the bottom, and thus render the water soft. The sulphate of lime or gypsum, or the chlorides or nitrate of lime, cannot be so easily eliminated, and water which contains considerable proportions of them is therefore called *permanently hard*.

Hard water also affects a good many colouring materials, and its most objectionable quality is that it forms deposits in steam-boilers, which are frequently very troublesome, and may be the primary cause of an explosion.

Wash-water which contains a sufficient quantity of lime to be hard, is unfit for a paper-mill.

Iron salts in contact with alkalies, lime, or soda, deliver their acid to the latter, and the iron precipitates as oxide or rust, colouring the pulp until it may be re-dissolved by sulphuric or other acids. Although the proportion of these salts be insignificant, the quantities of water used are so enormous that the total amount of iron is yet quite considerable. A 500-pound engine, for example, carries about 10,000 pounds of water, and if this contains only one-fiftieth of one per cent. of iron, there will be 2 pounds of it in the whole mass, and if this water is renewed five times during one washing operation, 10 pounds of iron will be brought in contact with the rags.

The soda, bleach-liquor, alum, or sulphuric acid absorbed or neutralised by these iron salts in the multitude of operations, in all of which water is an important factor, sum up to a large quantity in a short time. It is quite probable that the difference in the quantities of chemicals consumed for like operations on the same stock in different mills, may sometimes be traced to this source.

The presence of iron can easily be discovered by the addition of a solution of yellow prussiate of potash to the water; the iron salts will form with it Prussian blue.

The total amount of mineral impurities can be ascertained by evaporating carefully several gallons of water and weighing the residue.

For papers of a lower grade, such as wrapping, the preparation of which requires few chemicals, it is not a matter of vital importance; but as the colour of all white papers depends greatly on the purity of the wash-water, an abundance of pure, clear wash-water is one of the conditions of the successful manufacture of fine papers.

48. Sources of Wash-Water.—To determine which are the best sources for good wash-water, it is necessary to understand the manner of their formation.

The water which covers the surface of the earth changes its form constantly; it evaporates, and is taken up and carried away by the winds as vapour. The air is able to hold more water at a higher temperature than at a lower one; any cold wind will, therefore, cause some of it to drop in the form or rain or snow.

The water, on returning to the earth in this form, is as pure as it can be found in nature; it contains no foreign matters but the gases which it takes up in the atmosphere, and if it flows over hard, insoluble substances, such as rocks of granite or quartz, or over sand, it preserves this purity. Streams of this kind are very valuable, but unfortunately they can only be found in the mountains, in most cases too far from markets to be available.

The purity of all surface waters, such as creeks and rivers, depends entirely on the nature of the soil over which they themselves and their tributaries pass, and should be in every case investigated.

A great portion of the snow and rain filters though the ground and comes again to the surface in some lower places as springs, or gathers in large cavities below, to which access is had by means of wells.

Very often these underground lakes extend from high places to low ones, and are only prevented from rising to a uniform level by a stratum of water-tight materials. In boring an artesian well this stratum is pierced, and the water forces its way upward with a tendency to reach the level of the highest point of the body of water from which it comes. In some cases it rises through large pipes high enough to drive a water-wheel, while in others it hardly comes to the surface; sometimes the water obtained is very pure, and at others it is loaded with foreign elements.

Boring an artesian well in an untried place is like digging for hidden treasures,—a very uncertain undertaking.

One or more never-failing springs of pure water, furnishing a full supply, are very valuable in a good location.

Where good wells can easily be made, and where experience has shown that they keep their supply all the year, they are, if the water is chemically pure, often preferable to surface water. While the latter may require to be artificially filtered, the well water has been cleared by passing through the soil.

49. Systems of Distribution.—If the wash-water can be taken from a convenient place above the mill, so that it has fall enough to run directly into the engines, a considerable amount of power and some machinery, which would otherwise be necessary to force it up, will be saved.

Mills which are not so fortunate, must have receivers in some of the highest parts of their buildings, into which the water can be lifted by pumps, and from which it is distributed. These reservoirs must be water-tight, and, if of wood, should be circular; but iron is a better material for this purpose, as it does not shrink and open in the joints, like wood, when for a time empty.

If the pump has to stop for repairs, or from other

HENRY WATSON,

HIGH BRIDGE WORKS, NEWCASTLE-UPON-TYNE,

GENERAL MECHANICIAN.

SOLE MANUFACTURER OF HIS IMPROVED KNOTTER PLATES,

Also of improved Revolving and Jogging Strainers in Vats complete,

DOCTOR PLATES, BRASS AND COPPER ROLLS,

HYDRAULIC PRESSES AND PUMPS,

Gun Metal Cocks, Valves, Water and Steam Gauges, Hydraulic Rams, &c.

H. W. begs to intimate that to meet the increasing demand for his improved Strainer Plates he has just completed extensive additions to his Premises and Machinery, and is now in a position to execute all orders promptly, and on the most reasonable terms. The strictest attention will always be given to maintain in the highest degree the quality of the material and workmanship, keeping in view also the capability of re-closing after having been worn.

Rag Chopper, Wheel, 4 ft. diam., 1 ft. wide.   Finishing Calender. These rolls are made of refined Chilled Roll Metals.

BENTLEY & JACKSON,
ENGINEERS, IRONFOUNDERS, AND MACHINISTS,
BURY, NEAR MANCHESTER.

Makers of Paper Making Machinery, Millboard Machines, Paper Cutting Machines, Ripping and Winding Machines, for preparing paper for continuous printing presses.

Hydraulic Pumps and Presses.   Steam Engines and Boilers.

ESTIMATES ON APPLICATION.

IMPORTANT TRADE PUBLICATIONS.

Thirteenth Edition. Demy 8vo., 104 pages, Stiff Covers, Price 2s. 6d., or Post free for 30 Stamps,

THE PAPER MILLS DIRECTORY.

Price 2s. 6d., or Post free for Thirty Stamps,

THE PAPER STAINERS' DIRECTORY
OF GREAT BRITAIN.

The Seventh, a New, Corrected, and Enlarged Edition, Foolscap 8vo., price 3s. 6d., or Post free for 30 Stamps,

THE STATIONERS' HANDBOOK,
AND GUIDE TO THE PAPER TRADE.

Sixth Edition, Demy 8vo., 60 pages, Stiff Covers. Price 2s. 6d., or Post free for 30 Stamps,

THE CHEMICAL MANUFACTURER'S DIRECTORY.

Sold by KENT & CO., Paternoster Row.

Price 2s. 6d., or sent Post free for 30 Stamps,

A MAP OF THE PAPER MILLS OF ENGLAND.

Arranged by the Editor of "The Paper Mills Directory."

THE EDITOR, at Oxford Court, Cannon Street, London, E.C.

No. Price

THE ART

OF

PAPER-MAKING:

A Guide to the Theory and Practice

OF THE

MANUFACTURE OF PAPER.

COMPILED BY THE

EDITOR OF THE "PAPER MILLS DIRECTORY."

London:
SOLD BY UNWIN BROTHERS,
OXFORD COURT, CANNON STREET.

B. DONKIN & CO.'S SPHERICAL ROTARY RAG BOILER.

To CONTAIN FROM 20 TO 25 CWT. OF RAGS.

This Boiler being spherical, is twice as strong as a Cylindrical one of the same diameter and thickness. The Plates used are notwithstanding of the usual substance, thus rendering it perfectly safe, durable, and suitable for high pressure steam.

The Spherical shape has also another important advantage, viz. allowing all the rags to fall out by themselves, when the boiler is revolving with its cover off.

Inside the Boiler are Strainers to take off the dirt, and Lifters to agitate the rags during the process of either boiling or washing.

To avoid Cement or run lead Joints, the Gudgeons and the Boiler are turned true in a lathe to fit each other, the joints being simply made with red lead.

B. DONKIN & CO.,
ENGINEERS, MILLWRIGHTS, MACHINISTS, AND IRON FOUNDERS,
Blue Anchor Road, Bermondsey, London, S.E.

Manufacturers of Steam Engines, Turbines, Water Wheels, Flour Mills, &c., Paper Making Machines, and everything connected with Paper Mills, Rag Engines, Rag Cutters, Cutting Machines, Glazing and Hydraulic Presses, Pumps, Hoists, &c., Steel Bars and Plates, Wires, Felts, and Deckle Straps. Machines for Printing in two Colours. Gas and Water Valves.

causes, the mill has to be stopped also, unless the reservoir holds water enough to keep it going.

A strong foundation is at all times required for it, as the weight of water in even a small reservoir is considerable.

One large mill in America is supplied with a capacious settling pond or reservoir on the top of a hill, situated higher than any part of the machinery, and is filled weekly by means of a large force-pump, driven by water-power, with a sufficient supply of water for the six following days.

Two barrels of porous alum are emptied into this pond after it has been filled, in order to precipitate the impurities. The lime and iron, which may be contained in the water in the form of carbonates, will form sulphate of lime and sulphate of iron with the sulphuric acid of the alum, while its other component part, the alumina or clay, is set free and carries down mechanically some of the floating impurities.

50. **Quantity Required.**—It is impossible to calculate exactly the quantity of wash-water which is required for a paper-mill, but an estimate, on which the sizes of the pump and of the distributing pipes may be based, is indispensable. We will indicate how the data on which such an estimate may be based can be obtained, and coupled with experience it will be a sufficient guide.

The washing-engines consume by far the larger portion of all the wash-water, and we suppose, for example, that two of them, 15 feet long by $7\frac{1}{2}$ feet wide, holding each about 500 pounds of rags, are used in a mill of a capacity of about 3,000 pounds of white paper per day. It is true that they do not always wash at one and the same time, but it happens so sometimes, and we must be prepared in such cases to furnish enough water. If the receiver is large enough to hold all the surplus, so that no water need be wasted through the overflow, no power is lost, but if the receiver is small the pump has to furnish an excess, which during most of the time runs away.

The washers must be fed with as much water as they are able to discharge, and this will in most cases be amply done by a stream which will fill the empty tub in fifteen or twenty minutes. One of the engines taken for this example, holds 160 cubic feet = 1,200 gallons; two engines, therefore, require in fifteen minutes $2 \times 1,200 = 2,400$ gallons, or in one minute 160 gallons.

The quantity of water consumed in boiling, by the beaters, and by the paper-machine, is difficult to estimate, but it is seldom as large as that required by the washers, especially at the high rate estimated in this example. We are, therefore, pretty safe in taking double the quantity, calculated for the washing-engines, or $2 \times 160 = 320$ gallons per minute, as the whole supply needed.

An abundance of wash-water is one of the first conditions in the manufacture of paper, and it is therefore wise rather to waste power than to have an insufficient quantity of water. We have to make allowance for deficient work of the pump and leakage in many places, and may add one-fourth of the calculated number, or $\frac{320}{4} = 80$ gallons, and thus need a pump capable of throwing 400 gallons per minute.

The power which is necessary to raise such a quantity of water can easily be calculated. If it is, for instance, taken from a well, at a depth of 12 feet, and pumped into a receiver, the water surface of which is 38 feet above the ground, the total height through which it must be raised is $38 + 12 = 50$ feet. It takes as much power to raise the weight of 400 gallons or $400 \times 8\frac{1}{3} = 3,330$ pounds 50 feet high, as would be required to raise $3,330 \times 50$, or 166,500 pounds, one foot high. One horse-power is accepted as equal to the power which is necessary to raise 33,000 pounds one foot high in one minute, and our 166,500 are therefore equal to $\frac{166500}{33000} = 5$ horse-power actual work.

51. Pumps.—Piston or plunger-pumps, with suction or pressure-valves, although the oldest style, are even at present in many cases preferred to all others.

The valves must have time to open and close perfectly if a good result is expected, and their speed should therefore always be moderate.

The perfection of rotary pumps have been the study of numerous mechanics for many years, and the patents taken out for them would alone fill a good-sized book. These pumps work mostly without valves and run fast, but their speed, and with it the quantity of water thrown, can be considerably increased or reduced at will; they take little room, are operated by belts, can be easily set up, and require little care.

Rotary pumps, being generally less efficient as suction than as force-pumps, are usually set as close to the source of supply as possible; some working best, if the water is made to run into them without any suction. As force-pumps they are excellent, and some are even used for steam-fire engines.

The absence of valves makes it possible for pieces of

wood, rags, or other solid matters, which may accidentally be in the water, to pass through a rotary, while they would obstruct and perhaps damage a piston-pump.

The pressure of the atmosphere, at a low density, is equal to that of a column of water 30 feet high.

If it were possible to construct a pump so perfect, that it could withdraw all the air from the suction-pipe, thus creating an absolute vacuum, the water would rise in it to a height of about 30 feet, forced by the atmosphere outside; but if the water had to be raised but a trifle above this limit, the atmospheric pressure could not do it, and it would never reach the pump. The height to which any pump can raise water by suction is therefore always less than 30 feet.

If the suction-pipe is not perfectly air-tight, if the air enters through a small hole or crack, feeding the pump in the place of water, a vacuum cannot be created, and the pump will run empty. The hole may be small enough to be invisible to the eye; it may accidentally appear by the loosening of a joint, the opening of a sand-hole in the casting, or otherwise, and its discovery is often difficult.

It is most likely to be found by running the pump and creating suction, and then holding a burning light to all suspected places; wherever there is an opening the current will change the direction of the flame and draw it in.

This difficulty is a sufficient reason for such a disposition as will admit of as short a suction-pipe as possible.

There is another kind of pump which is frequently used in paper-mills, and recommends itself by reason of its simplicity.

It consists of a strong rubber belt carrying iron or copper buckets, which belt runs over two large flanged pulleys, one of which is located inside of a lower water-tank, while the other is fastened above the upper receiver.

The shaft of the upper roll is turned by a pulley and belt, and raises the buckets which have been filled in the lower tank. They travel up and down in close-fitting wooden spouts or troughs, and are emptied, as soon as they pass over the upper roll, into a receiving channel, which connects with the reservoir.

This pump is built on the same principle as a grain elevator, and works well where the water is not to be raised very high.

52. **Water-Power.**—The fall of a body of water, like that of any other substance, exercises a power, equal to its weight multiplied by the height of the fall, and if it is produced by the continuous flow of a stream, it can be utilized.

To find out how much power there is at a given point, it is necessary to establish, by a survey or by levelling, how many feet of fall can be obtained, and what quantity of water flows down the stream during a second or a minute.

It is comparatively easy to measure the fall of water, but more difficult to determine its quantity. The latter is obtained by multiplying the number of square feet of a vertical section of the stream with the velocity, or with the number of feet through which the water advances in one minute. If a race is at hand it may be used for this measurement; otherwise a part of the stream with as straight banks and as even a width as can be found, should be selected. Measure of it a certain length, say 100 feet, and in several points of this length the vertical section, viz. its width and medium depth. From these several ones the medium size of a section is calculated.

An instrument, resembling a wind-mill on a small scale, has been constructed for the measurement of the velocity of flowing waters, which, when set in a stream at any point, registers the number of revolutions of a fan, which enables us to obtain the speed.

If such an instrument is not to be had, a piece of light wood may be used in its place; it is simply thrown into the middle of the stream, at the point above where the measured part begins, while the time which it consumes in flowing down through the 100 feet length is observed with a watch. The number of feet made by it in one second is the velocity, and gives, multiplied by the number of square feet of the medium vertical section, the number of cubic feet of water which pass through the stream in one second.

If a precise calculation is to be made, the speed on the surface cannot be accepted as that of the whole body of water. The friction on the bottom and sides retards the motion, and must be taken into account. If we call the surface velocity per second, found with the floating wood V, the real velocity v of the stream is expressed, according to the best authorities, by the formula—

$$v = V \frac{7 \cdot 71 + V}{10 \cdot 25 + V}$$

If we have, for example, a race or stream of 20 feet medium width, 3 feet medium depth, with a surface velocity of 1 foot per second, the real velocity of the water will be :

$$1\frac{7{\cdot}71 + 1}{10{\cdot}25 + 1} = 0{\cdot}774 \text{ or } \frac{77}{100} \text{ feet,}$$

and the volume of water which flows through it during one second :

$$20 \times 3 \times \frac{77}{100} = 46{\cdot}2 \text{ cubic feet.}$$

One cubic foot of water weighs $62\frac{1}{2}$ pounds, and $46{\cdot}2$ cubic feet $= 2887{\cdot}50$ pounds. This quantity is available every second, or $60 \times 2887{\cdot}50 = 173{,}250$ pounds every minute. If there be a fall of 15 feet, the power is equal to the fall of $173{,}250 \times 15 = 2{,}598{,}750$ pounds through one foot per minute.

One-horse power being equal to the fall of 33,000 pounds through one foot height in one minute, the water power amounts to

$$\frac{2{,}598{,}750}{33{,}000} = 78{\cdot}\tfrac{7}{3} \text{ horses.}$$

53. Dams.—The proper construction of a dam depends so much upon the location and the material which can be had for it, that general rules cannot be given. If possible, it should be so situated that a large body of water can be accumulated behind it, which may be drawn upon in dry seasons. Lakes, as sources of supply, are excellent natural reservoirs.

If the dam and the mill are situated in a narrow valley where, in case of a flood, the water cannot spread over a large surface, but is stowed up high, the pressure sometimes becomes so strong that both dam and mill are swept away like chaff before the wind. Such sites on streams which are subjected to freshets, are dangerous, and require the construction of the best foundations for both mill and dam which human skill can devise.

The head-race conveys the water from the dam to the point where it begins to act on the mechanism. It maybe an open conduit or a closed pipe, or a combination of both. Economy of power requires, that it should be as large as possible; economy of first cost, that it should be as small as possible; the right mean must be chosen. The entrance of the water is regulated by the head gates at the starting-point, while the lower end pours it on a water-wheel or into the penstock of a turbine.

54. Water-Wheels.—The old-style water-wheel is always vertical, while its more recent competitor, the turbine, is horizontal.

Neither of them, nor any other hydraulic motor, is capable of returning the full actual power of the waterfall.

Vertical water-wheels are divided in undershot, breast, and overshot wheels, according to that part of the circumference where the water strikes them.

The overshot wheel returns the highest proportion of the actual power, while the results obtained with breast-wheels are less favourable the lower down on their outer circle the water is admitted. The overshot is therefore the only kind of vertical water-wheels which may compete with the turbine as a motor for paper-mills.

If the overshot water-wheel could be constructed so,

that the whole body of water would be carried by it from the highest to the lowest point of the available fall, and be there suddenly discharged, nearly the whole natural power would be realised, while, as it is, considerable losses will be suffered, some of which are here indicated.

The fall is represented by the distance between the surfaces of the water in the head-race and in the tail-race. The portion of this distance between the surface in the head-race and the wheel itself acts only by impulse, but not by weight.

Instead of the full outside diameter of the wheel, the distance between the point of gravity of the water in the uppermost and in the lowest bucket must be counted, as it is there where the weight of the water may be considered concentrated. Over half of the depth of two buckets, or the full depth of one bucket, is thus lost.

The wheel must hang free above the water in the tail-race, and the distance between wheel and water, representing a small portion of the fall, is also lost.

The buckets cannot be emptied *suddenly* at the lowest point; they require some—be it never so little—time for it, and must therefore begin to discharge at some height above a portion of the water, the weight of which will be lost for the balance of the fall.

If any water is left in the buckets on their ascent, its dead weight neutralises the same quantity of live weight on the descending side.

To all the losses just enumerated must be added those from friction and contraction in the races and wheel, from leakage, and bad construction.

To obtain the best possible effect, the wheel must be

built in such a manner that no part of it can leak or change position, and its top should be about 2¼ feet below the level of the water in the race. Wood will become warped, and will rot, and is therefore inferior to iron; but, if an all iron wheel is too expensive, the buckets alone may be of metal and the body of wood. Care must also be taken to provide an easy escape for the air contained in the buckets, through openings in the sole-plate. A guide-board, over which the water is made to flow in the right direction and to the right spot, is used to convey it from the head-race to the wheel.

The circumference of a water-wheel should, according to the best authorities, have—be its diameter large or small—a speed of not less than 4 feet and not more than 8 feet per second. A point on a large circle, if moving with the same speed as a point on a small one, requires more time to make a revolution than the latter; the larger a water-wheel is, the slower will therefore be the movement of its shaft.

Overshot water-wheels furnish from 60 to 80 per cent. of the natural water-power, according to the amount of care and skill applied to their construction and disposition.

55. Turbines.—The turbine is a horizontal water-wheel with vertical axis, consisting of a drum or annular passage with a set of vanes, curved like the surface of a screw, so that the water, after having exercised its impulse on them, will glance off with as little energy as possible. While the vertical wheel is moved principally by the weight of the water, the turbine is propelled by its impulse only, and to get the best effect, the water has to

be guided so that it will strike every part of the moving vanes as nearly as possible at a right angle.

The first turbine, invented by Fourneyron, was put in operation in the year 1827, at Pont-sur-l'Ognon, in France; this Fourneyron-wheel was then, and is yet, constructed of two concentric rings, both of which are open on their vertical sides, closed on their horizontal ones, and supplied with an equal number of vanes. The inner ring is stationary, receiving the water in the centre and acting as a guide only, while the outer one revolves and transmits the power, the water leaving at its circumference.

Jonval made the guide-wheel and the revolving-wheel of the same diameter, and placed them, one on the top of the other, in a vertical cylinder. He was thereby able to set the turbine at any height between the head and tail-races, with equally good results, provided that the lower portion, which may be called the suction-pipe or draft-tube, be less than 30 feet high above the level of the tail-race.

To prove this seemingly strange fact, we will take as an example a fall of 20 feet, with a turbine wheel incased in a water-tight cylinder or penstock of 20 feet height. If the wheel is placed at the lowest end, the water is forced through it with the pressure exercised by the 20 feet fall and by the pressure of the atmosphere—equal to 30 feet fall—or altogether by 50 feet. But the atmosphere has also free access to the water in the tail-race, and presses against the turbine with a force, also equal to 30 feet fall, which must be overcome. Deducting, therefore, these 30 feet from the 50 feet pressure from above, leaves only 20 feet, or the fall of the water, as active pressure.

If the same turbine is situated in the middle of the penstock, 10 feet from the surface of the water in either race, the water will be forced on to the turbine with a pressure which is equal to that of the column of water above the wheel, or to 10 feet in addition to the atmospheric pressure, or altogether to $10 + 30 = 40$ feet. But before the atmosphere can in this case exercise any pressure against the wheel from the lower side, it must first overcome the column of 10 feet in the draught-tube below the wheel, and is therefore reduced to 30 less 10 or to 20 feet. The difference between the pressure above and the resistance below is therefore 40 less 20, or 20 feet, as before.

The draft-tube below a turbine acts like the suction-pipe of a pump; if air finds admittance into it, power equal to a fall of the same height is lost.

There is always danger that an opening may be caused by some accident, by faults in the material, or by wear and tear; and even the most insignificant holes, which can hardly be seen, must cause some loss of power. In most cases these pipes are located where it is very difficult to examine them, and much valuable power is often wasted before the faulty spot can be discovered. Suction-pipes or draught-tubes for turbine-wheels should therefore be dispensed with, except in cases where for some reason the wheels cannot be set low enough to do without them.

Some turbines which have within late years been much used, are remarkable especially for simplicity, and consequent cheapness. Every opening is provided with a gate, formed and fastened in such a manner, that it serves at the same time as a guide to the entering water. All these gates can be opened and closed by one common rod, with which they are connected by levers or gearings. The water enters on the outer circle and escapes inside, but, instead of running away at once, it passes another set of differently curved vanes, intended to absorb any power which may have been left in it.

These horizontal wheels always run fast, and, like the vertical ones, the more so the smaller they are. They must be made of metal, because wood would take up too much space and could not be moulded into the required shape. The penstock and its enlargement, in which the wheels run, or the casing, is often built of wood; but it is advisable to construct it of iron at all times, especially when it is exposed to the pressure of a high fall. The first expense will be larger, but the almost inevitable escape of water by leakage and the constant repairs of a wooden structure will be avoided.

Turbines lose power, like overshot wheels, by friction and contraction, and because they cannot be constructed sufficiently perfect to absorb all the power. They return from 60 to 80 per cent. of the natural power; 75 per cent. may be considered a very good result, while 80 is only obtained in exceptional cases.

56. Comparative Advantages of Overshot and Turbine Wheels.—Overshot wheels may be used when the available waterfall is such that their diameter will be reasonably large—not below 12 and not above 25 feet; if the diameter would have to be beyond these limits, a turbine would be preferable.

Turbines, being submerged in water, are never frozen up, and although their power will be reduced by backwater in proportion to the diminution of the fall, they cannot be stopped by it like vertical water-wheels.

Large gearing is required to produce from the slow motion of a vertical water-wheel the high speed required by the line-shaft of a paper-mill, while one pair of comparatively small bevel-wheels only is necessary with the fast-running turbine.

Wherever the supply of water is either abundant or steady, the turbine will give a regular speed and good effect. But it returns the highest percentage of power only with the quantity for which it has been constructed; and if the supply should decrease and the water fall in the race, the power produced would not only be lessened in proportion to the loss of height and volume, but the percentage obtained from the remaining waterfall would be decreased fearfully. It is therefore imperative to keep the head-race full all the time, and rather to stop and accumulate water than to use it as it comes, in inadequate quantity.

In cases where the water-supply is often insufficient, and where no very large reservoir or pond is at hand, an overshot wheel may be preferable to a turbine, because it will give larger proportions of the natural power with decreasing quantities of water.

57. Steam-Boilers.—Steam-boilers form a very important part of the equipment of a paper-mill, and yet they are sometimes treated with a negligence which is criminal, from the danger to which every person within their reach is exposed.

The consumption of fuel depends so much on the nature, construction, and treatment of the boilers, and is very often so heavy an item in the list of expenses, that too much care cannot be bestowed upon their selection and management.

58. **Heating-Surface.** — The gases resulting from the combustion of coal on the grate have, on starting, a temperature of about 2,400° Fahrenheit, and should transfer as much of this heat as is possible to the water contained in the boiler. To create a good draught in the chimney, a temperature of about 600° above that of the outside air is required; 1.800°, or three-quarters of all the heat created, is therefore available. The boiler must be constructed with a view to absorbing all of these 1,800°.

The rapidity with which heat is transferred from one body to another is proportionate to the difference of temperature between them; the gases of combustion should therefore be conducted in such a course along the boiler that this difference will be at all points as large, and as uniformly the same, as possible—an object which will be best attained, if the gases are brought in contact with the coldest part just before escaping into the smoke-stack, and with the hottest part immediately after leaving the furnace.

The water being fed in at the lowest point of the boiler, it follows from the rule just given, that the gases should pass first along the upper hottest portions, descend gradually to the colder lower ones, and leave near the entrance of the feed-water.

That portion of the shell which is covered with water inside and exposed to fire or hot gases outside, is the heating-surface. The capacity of a boiler for raising steam is directly proportionate to this heating-surface, the size of which, expressed in square feet, indicates—if the boiler is otherwise correctly constructed and supplied with a sufficient grate-surface—its value better than a certain number of horse-power.

The size of the heating surface which is to represent one-horse power has not been established by the trade, and the seller is therefore at liberty to represent the boiler as powerful as his, sometimes elastic, conscience will admit.

Fifteen feet heating-surface at least should be allowed for one horse-power, although one-half of it may by forced firing be made to evaporate the same quantity of water.

In calculating the power of a boiler, it is to be considered, that the lower, nearly horizontal part of internal flues or tubes, owing to the difficulty with which bubbles of steam escape from under them, are found to be much less effective than the lateral and upper surfaces. To obtain therefore the real useful heating-surface, we have to deduct nearly one-third from the total one of the tubes or flues. On an average, the effective heating-surface is from $\frac{3}{4}$ to $\frac{2}{3}$ of the total heating-surface.

It is always safer to buy a boiler of a fixed amount of heating-surface than of a number of undefined horse-power.

59. **Combustion.**—The useful parts of all fuel consist of the element carbon, which constitutes the solid parts, and of combinations of hydrogen and carbon in the forms of olefiant gas, pitch, tar, naphtha, &c. Both these elements, carbon and hydrogen, are, through the process of com-

bustion, combined in gaseous form with the oxygen of the air, and escape as carbonic acid (CO_2) and water (HO).

If the disengaged carbon is chilled by a cold draught or otherwise below the temperature of ignition, before coming in contact with oxygen, it constitutes, while floating in the gas, *smoke*, and when deposited on solid bodies, *soot*. But, if the disengaged carbon is maintained at the temperature of ignition, and supplied with oxygen sufficient for its combustion, it burns, while floating in the inflammable gas, with a red, yellow, or white flame.

If the boiler, or more properly its heating-surface, is small, and the firing hurried in order to produce enough steam, the combustion must be imperfect, and a loss of fuel will be the consequence. A large heating-surface may, compared with a small one, save in one year the cost of a boiler in fuel.

It has been found by experiments and calculation that it takes about 24 pounds of air to furnish enough oxygen for the combustion of one pound of coal, and to dilute the gases properly.

It seems evident that such a large amount of air as 24 pounds, equal to 650 cubic feet, for one pound of fuel, cannot be introduced within the furnace without artificial means or draft.

60. Draught.—This draught is usually produced by a chimney, and sometimes by a fan or other blowing machine.

The gases inside of a chimney are expanded by heat, and therefore lighter than those outside, and the draught is proportionate to the difference in weight between the column of gases inside and that of an equal column or volume of air outside. The efficiency of a chimney depends, therefore, principally on the height of its crown above the fire-grate. Several formulæ have been proposed by which it is to be calculated, but local experience has usually the deciding voice.

It is, however, advisable to build the smoke-stack high enough to answer, not only present demands, but also increased ones which may be made in the future.

Small pieces of coal which have escaped through the chimney, can frequently be found in the screens, and sometimes in the paper, but if the stack is very high, the smoke will be carried off to a distance before its floating solid parts can reach the ground—a great advantage, especially where soft or bituminous coal is used.

As a rule which will answer in most cases, a chimney may be built to a height of twenty-five times its inside diameter or width in the clear. The area of the width of a chimney can be made 0·16, or $\frac{1}{6}$ the area of the fire-grate, if the latter is of the ordinary construction, or equal to the sum total of the area of the flues in any one place on the course of the hot gases.

The hot gases cool off, contract, and consequently require less room as they ascend through the chimney, and many scientific writers therefore recommend building the stacks conical or pyramidal inside and outside, that is, narrower towards the top than at the bottom.

It has, however, been lately found that smoke-stacks, constructed as inverted cones, or wider at the top than at the bottom, give a better draft. This contradiction of the long-followed theory is in practical use on most locomotives and in the brick chimneys of numerous factories.

and it may be stated, that such stacks are much lower than anybody would dare to build them on the contraction plan, and that they generally give satisfaction.

They are perfectly perpendicular outside, and are made funnel-shaped inside by means of a double wall. The outside wall may, for instance, start at the bottom with a full brick of 9 inches, and run out at the crown with one-half brick, or $4\frac{1}{2}$ inches, while the inside wall is only half a brick, or $4\frac{1}{2}$ inches, thick, parallel with the outside one, connected with it at intervals by a brick or *binder*, but leaving a few inches, distance between the two. This inside lining is only carried up for a part of the height, and then broken off, thus leaving the top considerably wider than the lower part.

The foundation of a chimney should be as solid as a rock, as it has to sustain the enormous weight of the bricks which are piled upon it. The slightest sinking of a part of the foundation may cause the top to lean over, and perhaps to fall.

Access must be provided to the inside of the stack, through an iron door or through an arch near the bottom, for the removal of the ashes which will gather there in the course of time.

The hot gases always carry some fine dust or coal along, and deposit them on parts of the boiler over which they pass; the portion of the heating-surface thus covered is ineffective, and it is therefore imperative that the flues and boiler-surfaces should be frequently cleaned, and doors must be provided for that purpose.

61. Grate-Surface and Firing.—The ordinary rate of combustion for factory boilers is, according to Rankine, from 12 to 16 pounds of coal per hour on every square foot of grate-surface, the size of which may be approximately determined from this. The same writer not only recommends a sufficiently large grate-surface, but warns us against any increase of its size beyond the prescribed limits, because too much air would then be admitted, and absorb a great deal of the heat without being of any use.

The admission of air can, however, be well regulated by the damper, and it cannot be conceived why a large grate-surface, which permits slow combustion with little draft and a light covering of fuel, should be objectionable.

We have lately seen a system of boilers which confirms this opinion. They are twin boilers, consisting of a lower cylinder, fitted with 2 or 3-inch flues, and an upper plain cylinder. They are not walled-in in the ordinary manner; the lower cylinder, resting with its ends only in the two short walls, hangs free, so that the fire can play all around it. The upper cylinder rests above the lower one in the same two end walls, and the long side walls are arched up, so that they join it all along on both sides. This brings the upper half, or its steam room, beyond the reach of the flames, while the lower half is open to the fire, like the lower flue-cylinder, from which it is not separated by any arches or partitions. The grate occupies the whole space between the four walls under the boiler, and is three or four times as large as usual. The fire-doors are on the long side of the boiler, and as numerous as the room permits. When two such boilers are required, they are walled-in between the same four walls and above one common grate, without any separating walls. The grate being very wide, it is then necessary to have fire-doors on both long sides. The fire or hot gases, after having passed around the lower cylinder and below the upper one, are conducted to the outside at the end, which in ordinary boilers contains the fire-doors, and descend from there through iron conduits into the flues of the lower cylinder, through which they pass to the stack. The grate should be at a convenient height above the ground for the work of the fireman, and as the lowest part of the boiler must be at some distance above the fuel, the whole structure becomes necessarily very high. The large grate-surface is slightly covered with coal, which is thrown in through each door in regular succession, and thus kept uniformly spread. Very little draught is required, and the combustion seems to be perfect.

We have been informed that these boilers furnish in one case, with 7 tons of coal, as much steam as others, which had been previously used, would produce with 10 tons; and that in the other the consumption of coal had been reduced by their introduction from 20 to 9 tons for exactly the same work.

If a boiler is provided with only a small grate-surface, the necessary amount of combustion of fuel cannot take place without a strong draught, or in other words, the air must be forced through the coal and along the boiler with great rapidity, sometimes so fast that it can neither become well heated or thoroughly deprived of its oxygen. The necessity for a strong draught indicates, therefore, that the grate-surface is not sufficient; and it would probably result in a great saving of fuel for many boilers if their fire-places were extended and their draughts reduced.

Smoke is not only a nuisance, but also a loss of as much unburnt coal; it escapes in thick clouds when the fire-doors are opened and fresh, especially finely divided, coal is thrown in. The cold air reduces the temperature, while the draught carries off small particles of coal untouched.

A simple way to prevent this loss, to some extent, which is applicable to any common steam-boiler, consists in the division of the grate into two separate parts by a brick wall in the middle parallel with the grate-bars. Each division has a separate door, and while the heat is greatest in one of them, fresh coal is thrown into the other; the unconsumed coal which is carried off by the draught mingles with the flames from the hot division, and is thus burnt up. The sudden changes of temperature which are so injurious to the boilers, caused by the opening of the fire-door, are thereby also in a measure prevented.

A constant stream of fresh air is often conducted into the fire-hearth through channels in the brickwork, wherein the air on its passage is partly heated.

Ordinary grates are composed of straight, narrow cast-iron bars, laid alongside of one another, and leaving, for a coal-fire, openings enough between them to amount altogether to one-fourth of the area of the grate-surface.

These bars when heated expand, and become frequently warped into such crooked forms, that they have to be replaced by new ones. Many patent grate-bars have been constructed with a view to prevent any change of their outer form, or of the open space for the admittance of air, by providing room for expansion and contraction in the bars themselves. Some of them are quite successful, and not only save coal, but also last much longer than the common bars.

It is a mistake to suppose that the sprinkling of water over coal will improve combustion; the water must be evaporated and transformed into steam, absorbing thus a great deal of heat at the expense of the fuel.

The coal must be spread on the grates uniformly, and not too thickly, and if stirred at all, it is to be done from below.

A good fireman can economise more than all the inventions which have been made for this purpose are able to do, while a careless or ignorant one may waste many times the amount of his wages in fuel.

Several years ago prizes were offered to such firemen as should prove themselves most efficient at a competitive trial, at Mühlhausen, then in France, now belonging to Germany. Fourty-four offered themselves, and the best eighteen were selected from the number; each one fired up during ten hours with the same boiler, fuel, &c. and it was found that the best fireman could evaporate nearly twice as much water with the same amount of coal under exactly the same circumstances as the worst one. To appreciate this result, it must be considered that only experienced men offered themselves for the trial, and that only one-half of these were admitted.

62. Construction of Steam Boilers and Test.—The designs for the construction of steam-boilers are so numerous that a description cannot be attempted in this work, but a few words may be said which will apply to them all.

It has been found by experience that a thickness of

⅜ inch is the most favourable to sound riveting and caulking of boiler-plates, and they are therefore seldom used much thicker or thinner.

If a cylindrical boiler is required to endure an unusually high pressure, the necessary increase of strength must be attained, not by increased thickness of the plates, but by diminution of the diameter.

The flat ends of cylindrical boilers are given about one and a half the thickness of the cylindrical portions; cast-iron should not be used for them nor for any other part of the shell. These flat ends are usually connected with each other by longitudinal stays, and sometimes with the cylinder by means of angle iron, but such rings are liable to split at the angles, and it is therefore preferable to bend the edges of the flat ends and rivet them to the cylinders.

Plates which overlap one another should have the overlapping joints facing upwards, on the side next to the water, that they may not intercept bubbles of steam on their way upwards. The joints in horizontal flues should be placed so that they do not oppose the current of the gases. Those parts of boilers which are exposed to more severe or more irregular strains than the rest, or to a more intense heat, should be made of the finest iron.

Lately there have been some steam-boilers built of steel plates. As they have about one-half more tenacity than iron ones, they can be made lighter, and it is not unlikely that steel boilers will take the place of iron ones at some future day.

In paper-mills the demands for steam are often so sudden and large, that some of the water will be carried along mechanically through the violence of the motion, unless a large storeroom for both water and steam, but especially for the latter, is provided. A steady pressure is particularly required for the paper-machine, and large boilers, with plenty of steam room, will be more apt to furnish it than small ones.

Every boiler, before being put into operation or walled-in, should be tested with a hydraulic pressure twice as great as the highest steam-pressure allowed to it. Water-pressure is used on account of the absence of danger, in case any part of the boiler should give way.

63. **Feed-Water.**—If possible, the boilers should be fed with hot water; not only because as much heat as it contains is directly saved, but also for the reason that the injection of cold water chills and may injure the hot plates. There are plenty of sources in a paper-mill from which hot water may be obtained, but the dryers of the paper-machine are the principal ones. The condensed steam is conducted through a pipe into a reservoir or tub, and if too hot to be pumped, it is therein mixed with fresh water and forced into the boiler by the feed-water. These pumps refuse to work with highly heated or boiling water, and large quantities of escaped steam from steam-engines, by which the temperature of the feed-water might be raised to the boiling-point, are often allowed to blow out into the open air, because they would heat the water too much.

Such steam can, however, be utilised by being conducted through long coils of pipe, fastened in an upright steam tight cylinder of boiler iron, interposed between the pump and the boiler. The feed-pump forces the hot water first into this cylinder or heater, where it acquires a very high temperature in contact with the steam coil, and thence through the usual check-valve into the boiler. The feed-water does not pass through any pump after it has left the heater, and can therefore be raised to any temperature.

It is of the greatest importance that the water in the boiler should be kept at the right height all the time, and since the regular feed-pump may get out of order, there should be a second one, or some other means provided, by which the supply can be kept up in such a case. Giffard's injector or steam-pumps may be recommended for this purpose on account of their independence of any gearings or motors, but as all of them consume steam, and some will not feed with hot water, the regular feed-pump is always to be used in preference.

A valve must be provided at the lowest point of every boiler, through which it can be emptied; this valve or blow-off cock must be frequently opened to enable the deposits of salts or mineral matters to escape before they have had time to solidify. According to the quality of the feed-water this has to be done several times, or only once a day, and the boilers should in all cases be emptied entirely at regular intervals.

Some substances occurring in feed-water seem to stick so closely to iron that they cannot be removed by blowing off water; they must be chemically dissolved, and numerous powders and liquids are sold at high prices for this purpose, but the introduction of one or more gallons of common coal oil or petroleum will answer in many cases just as well or better.

64. Explosions.—We quote here from remarks made on this subject by W. T. Macquorn Rankine, Professor of Engineering and Mechanics in the University of Glasgow, in his *Manual of the Steam-Engine and other Prime Movers* —a work which has furnished much data for this chapter:—

" Explosions result :

" I. *From original weakness.* This cause is to be obviated by due attention to the laws of the strength of materials in the designing and construction of the boiler, and by testing it properly before it is subjected to steam-pressure.

" II. *From weakness produced by gradual corrosion of the material of which the boiler is made.* This is to be obviated by frequent and careful inspection of the boiler, and especially of the parts exposed to the direct action of the fire.

" III. *From wilful or accidental obstruction or overloading of the safety-valve.* This is to be obviated by so constructing safety-valves as to be incapable of accidental obstruction, and by placing at least one safety-valve on each boiler beyond the control of the fireman.

" IV. *From the sudden production of steam of a pressure greater than the boiler can bear, in a quantity greater than the safety-valve can discharge.* There is much difference of opinion as to some points of detail in the manner in which this phenomenon is produced, but there can be no doubt that its primary causes are, first, the overheating of a portion of the plates of the boiler (being in most cases that portion called the ' crown of the furnace,' which is directly over the fire), so that a store of heat is accumulated ; and, secondly, the sudden contact of much overheated plates with water, so that the heat stored is suddenly expended in the production of a large quantity of steam at a high pressure.

"Some engineers hold that no portion of the plates can thus become overheated, unless the level of the surface of the water sinks so low as to leave that portion of the plates above it uncovered; others maintain with Mr. Boutigny, that when a metallic surface is heated above a certain elevated temperature, water is prevented from actually touching it, either by a direct repulsion or by a film or layer of very dense vapour, and that when this has once taken place, the plate being left dry may go on accumulating heat and rising in temperature for an indefinite time, until some agitation or the introduction of cold water shall produce contact between the water and the plate, and bring about an explosion. All authorities, however, are agreed that explosions of this class are to be prevented by the following means:—

1. "By avoiding the forcing of the fires, which makes the boiler produce steam faster than the rate suited to its size and surface.

2. "By the regular, constant, and sufficient supply of feed-water, whether regulated by a self-acting apparatus or the attention of the engineman to the water-gauge; and

3. "Should the plates have become actually overheated, by abstaining from the sudden introduction of feed-water (which would inevitably produce an explosion) and by drawing or extinguishing the fires, and blowing off both the steam and the water from the boiler."

To overcome this difficulty several inventors have constructed boilers which are combinations of cylinder-boilers with systems of wrought-iron tubes. But whatever device may be used, and however excellent it may be in other respects, a steam generator cannot be considered a safety-boiler, if fire or hot gases are allowed to come in contact with a cylinder of more than 4 inches diameter, which contains water and forms a part of the boiler.

65. **Steam-Engines. Expansion.**—The steam-engine has been much improved since its invention about 100 years ago, but the fundamental principles governing its construction are yet the same as laid down by Watt.

The power of the steam-engine is derived from the alternate action of the steam upon the two sides of a piston, which is thus moved from one end of a cylinder to the other, the reciprocating motion being changed into a rotary one by means of a crank.

We suppose the piston of a high pressure engine, for an example, to have arrived at one end of its course, and to be on the point of starting to return, forced by steam admitted into the narrow space behind it, while the empty cylinder in front communicates with the open air. If the steam is of 60 pounds, equal to 4 atmospheres over-pressure, its real pressure will be 75 pounds, or 5 atmospheres, which, being opposed by one atmosphere, or 15 pounds only on the other side, pushes the piston forward with 60 pounds to every square inch of its surface.

If fresh steam is admitted constantly during the whole course of the piston, the largest amount of power of which the cylinder is capable will be produced, but the steam leaves the engine with nearly its full pressure.

If we take, for a second example, a cylinder of the same diameter, but of twice its length, and admit only the same amount of steam for each stroke, as in the first example, we find that, after the fresh steam has been shut off, the piston is yet moved forward to the other end by the steam, which filled one-half of the cylinder.

This second half of the movement of the piston is produced by expansion, and through it, if extended far enough, can the pressure of the steam be utilised and reduced until it is nearly equal to that of the atmosphere, or 15 pounds.

It is evident that all the power produced by expansion in the second example, with the same quantity of steam as was used for the first example, is clear gain as compared with the latter, and though this is not exactly so in practice, it yet explains the economy of expansion. These examples show also, that a larger cylinder is capable of producing the same power with less steam than a small one, or that engines of ample capacity are the most economical, if provided with proper arrangements for expansion.

66. **Condensation.** — Another way of increasing the power of an engine is to reduce the counter-pressure by the creation of a vacuum.

The steam, instead of escaping into the air, is for this purpose conducted into an apparatus, where it is suddenly condensed by contact with cold water finely divided by a sprinkler. The water used for this condensation is thereby highly heated, the counter-pressure reduced considerably below 15 pounds, and the power exercised on the piston increased as much.

The use of these condensers thus enables an engine to work with very low steam-pressure, and reduces the danger to which high-pressure in the boilers exposes it.

67. Different Systems of Engines, and Utilisation of Escaping Steam.—All engines may be divided into—
1. Non-condensing or high-pressure engines;
2. Condensing engines.

High-pressure engines are simple in construction, easily managed, and therefore generally used whenever steam is only an auxiliary power, to be stopped and started according to the state of the water-power. The valve which admits the steam is usually regulated by a governor, which is set in motion by the line-shaft. If the shaft turns too fast or too slow, the governor closes or opens the valve, letting in less or more steam, and increasing or decreasing the expansion.

The escaping steam should never be allowed to waste directly into the air; it can be made useful by passing through a coil or other system of pipes immersed in water, or it may be conducted through large pipes, and heat the building, or into the mixing-pans to boil the liquors.

In some mills, which run by steam-power altogether, the escaping steam is thoroughly used up for boiling waste-paper in tubs, and by heating the mill and the feed-water for the boilers. One mill, which from waste-paper and with steam-power only, produces 5,000 pounds of good printing paper per day, economises so well that direct steam is not used anywhere except in the steam-engines. It has been stated that the establishment consumes only 1,500 pounds of coal per day over and above the amount which would be required if the power were furnished by water.

The dryers form a natural condenser for the high-pressure engines which are used for driving paper-machines.

Condensing engines require large quantities of water, are complicated and expensive, but they furnish more power with the same amount of fuel than any other kind. This is especially the case when high-pressure steam first acts in a small cylinder, from which it passes into a larger one, where it propels the piston by expansion and condensation.

The escaping steam can, however, be so well utilised in many paper-mills that the simpler high-pressure engines answer often as well as more complicated ones.

It is indifferent whether an upright or horizontal engine is selected, provided it be a good one, fastened on a solid foundation.

68. **Power of Engines.**—The statement that a steam-engine gives a certain amount of horse-power must be made in connection with that of the dimensions of the cylinder, steam-pressure, speed, expansion, &c. if it is to be of any value. A steam-engine will give nearly double power if its speed is doubled. Fast-running engines not only wear out soon, but get more easily heated and out of order: and while it is the seller's interest to speed them high and represent them more powerful, the purchaser's is just the reverse. Steam-engines should therefore be purchased according to their size, but not by the horse-power.

The power of an engine is the total mean pressure on

HENRY WATSON,

HIGH BRIDGE WORKS, NEWCASTLE-UPON-TYNE,

GENERAL MECHANICIAN.

SOLE MANUFACTURER OF HIS IMPROVED KNOTTER PLATES,

Also of improved Revolving and Jogging Strainers in Vats complete,

DOCTOR PLATES, BRASS AND COPPER ROLLS,

HYDRAULIC PRESSES AND PUMPS,

Gun Metal Cocks, Valves, Water and Steam Gauges, Hydraulic Rams, &c.

H. W. begs to intimate that to meet the increasing demand for his improved Strainer Plates he has just completed extensive additions to his Premises and Machinery, and is now in a position to execute all orders promptly, and on the most reasonable terms. The strictest attention will always be given to maintain in the highest degree the quality of the material and workmanship, keeping in view also the capability of re-closing after having been worn.

Rag Chopper, Wheel, 4 ft. diam., 1 ft. wide.

Finishing Calender. These rolls are made of refined Chilled Roll Metals.

BENTLEY & JACKSON,
ENGINEERS, IRONFOUNDERS, AND MACHINISTS,
BURY, NEAR MANCHESTER.

Makers of Paper Making Machinery, Millboard Machines, Paper Cutting Machines, Ripping and Winding Machines, for preparing paper for continuous printing presses.

Hydraulic Pumps and Presses. Steam Engines and Boilers.

ESTIMATES ON APPLICATION.

IMPORTANT TRADE PUBLICATIONS.

Thirteenth Edition. Demy 8vo, 104 pages, Stiff Covers, Price 2s. 6d., or Post free for 30 Stamps,

THE PAPER MILLS DIRECTORY.

Price 2s. 6d., or Post free for Thirty Stamps,

THE PAPER STAINERS' DIRECTORY
OF GREAT BRITAIN.

The Seventh, a New, Corrected, and Enlarged Edition, Foolscap 8vo., price 3s. 6d., or Post free for 30 Stamps,

THE STATIONERS' HANDBOOK,
AND GUIDE TO THE PAPER TRADE.

Sixth Edition, Demy 8vo., 60 pages, Stiff Covers. Price 2s. 6d., or Post free for 30 Stamps,

THE CHEMICAL MANUFACTURER'S DIRECTORY.

Sold by KENT & CO., Paternoster Row.

Price 2s. 6d., or sent Post free for 30 Stamps,

A MAP OF THE PAPER MILLS OF ENGLAND.

Arranged by the Editor of "The Paper Mills Directory."

THE EDITOR, at Oxford Court, Cannon Street, London, E.C.

the surface of the piston, less the pressure against it in pounds, multiplied with the velocity of the piston in feet per minute. This product must be divided by 33,000 if the theoretical horse-power is to be obtained; and from it we have to deduct for condensing engines 25 per cent. and for high-pressure engines 13·1 per cent. loss from friction and pumps, in order to find the actual horse-power. While the steam is expanding in the cylinder, its pressure decreases constantly, and to make an exact calculation its mean pressure must be determined.

If we have, for example, a high-pressure engine of 15 inches diameter of piston, 3 feet stroke, 60 pounds of steam, 50 revolutions per minute, and an expansion of $\frac{1}{2}$, the steam acts on a piston surface of—

$$\frac{15 \times 15 \times 3\cdot14}{4} = 176\cdot6 \text{ square inches,}$$

and the mean pressure on the piston for a steam-pressure of 60 pounds above the atmosphere, or for $60+15=75$ pounds, is 63·487 pounds. The atmospheric counter-pressure, which is equal to 15 pounds to the square inch, must be deducted from these 63·487, and leaves 48·487 pounds as the available pressure. The theoretical power is therefore—

$$\frac{\underset{\text{Square inches piston surface.}}{176\cdot6} \times \underset{\text{Pressure per square inch.}}{48\cdot5} \times \underset{\text{Feet stroke.}}{3} \times \underset{\text{Revolutions per minute.}}{50} \times \underset{\text{Strokes during one revolution.}}{2}}{\underset{\text{One horse-power.}}{33,000}} = 77\cdot85 \text{ horse-power.}$$

and deducting from the theoretical power

13·1 per cent., = 10·20 loss,

we obtain

the actual horse-power = 67·65

M

69. Losses of Power.—We have supposed the pressure to be 60 pounds, as indicated by the gauge; but if this is the pressure in the boilers, the steam will have lost a considerable portion before it reaches the engine; even if the conducting-pipes are short and well covered, the difference may amount to many pounds.

The cylinders are usually not so well covered as they should be, and lose pressure by the radiation of heat.

Water takes up frequently a part of the room which should be occupied by steam; and if the piston does not fit to a nicety, fresh steam escapes between it and the cylinder.

The counter-pressure is always higher than that of the atmosphere, or 15 pounds, as we have supposed it to be; the friction of the steam in the waste-pipes consumes power, and allowance must be made for the loss of fresh steam, which fills the channels and the space between the piston and the heads of the cylinder.

If the escaping steam is utilised, as it ought to be, by passing through pipes, surrounded by air, or water, or other liquids, or by direct introduction into the latter, the counter-pressure will be thereby considerably increased.

The total amount of loss from these sources is very different, according to the construction, disposition, and management of the steam-power; it can hardly be calculated, but can only be found by experience; a sufficient allowance should be made for it, so that the engine need not be forced.

The pipe which conducts fresh steam to the engine should not have less inside diameter than one-quarter of the diameter of the piston, and the escape or waste-pipe is to be as large as possible, but not less than one and a half times as large as the steam-pipe which connects with the boilers.

The cylinder is to be provided with small pipes fastened to the lowest points of the heads, through which the condensed water can be blown off outside of the room.

70. Disposition and Management. — Steam-engines of very regular speed are required in paper-mills, and especially for the paper-machines; they should be well built, and provided with sufficiently large fly-wheels and good governors.

They must be mounted on solid frames in such a manner that none of their parts can deviate in the slightest degree from their relative situations. Too much attention cannot be given to this point, as well as to solid, heavy foundations.

The connection between the steam-engine and shafting is now generally made by means of belts, either from separate pulleys or directly from the fly-wheel.

If the steam-power is only used to supply the deficiencies of a water-power, steam-engine and water-wheel may both drive the same shaft. The water-wheel gate must be opened sufficiently to admit all the water furnished by the steam; while the engine, which drives the same line shaft with a belt, is regulated by the governor, so as to furnish the balance of the required power, whatever it may be.

No greater mistake can be made than to give charge of a steam-engine to a cheap but incompetent man. Not only will the engine itself be ruined, but it will not furnish as much power from a certain quantity of steam as it should, and consume in wasted fuel many times the wages of a good engineer.

The best available man should be selected for this purpose, one who is desirous of instructing himself, and who takes pride in the good performance and clean condition of his engine.

71. The Fourdrinier Paper-Machine.—The paper-machine has not, like the steam-engine or locomotive, come forth in a comparatively finished state from the brain of one favoured man. It has required the life-long labour of many talented mechanics and manufacturers to change the ancient paper-maker's vat and form into the complicated mechanism, which now, on exactly the same principles, produces millions, where only thousands of pounds could be turned out formerly.

The first patent was taken out for a machine, making endless paper, by Louis Robert, in France, 1799, and he was awarded a prize of 8,000 francs by the Government. The troubles in which France was involved at that time caused him to bring his model to London, where the Messrs. Fourdrinier took it up. After spending a fortune and many years of work they succeeded in building a paper-machine which worked tolerably well. In 1807 they stated before Parliament that they had expended £60,000 to overcome the difficulties which they had encountered, and that they did not receive much encouragement from paper manufacturers.

They were never rewarded for their labours in a pecuniary way, and they certainly well deserve to be immortalised in the name of the present Fourdrinier machine.

We do not presume to exhaust this subject in all its bearings, but will try to point out the principles which should govern the construction as well as the management of a paper-machine, and give descriptions which may be of practical value to at least a portion of our readers.

The operations performed under this head may be classed as follows :

I. Regulating and diluting the pulp to its proper state by means of the regulating box, fan-pump, &c.

II. Freeing the stuff from impurities or substances other than single fibres.—This is done by sand-grates and screens, or pulp-dressers.

III. Forming the paper.—This is the most important part, and treats of the wire-cloth and its attachments.

IV. Forcing water out by pressure.—The presses.

V. Heating the paper until all the water has been evaporated.—Dryers.

VI. Polishing the surface.—Calenders.

VII. Winding up.—Reels.

VIII. Trimming and Cutters.—Cutters.

72. **Regulating and Diluting the Pulp.**—The regulating box is usually of wood, about $1\frac{1}{2}$ by 2 feet, and $1\frac{1}{2}$ to 2 feet high. It is divided by two upright partitions into three compartments, the middle of which receives the pulp, and empties it through copper gates into the two side compartments. One of the latter empties through a 3 to 5 inch copper pipe into the mixing box, or into the fan-pump, while the other is connected by a spout with the stuff-chest.

The pump always throws more pulp than is used, in order to make sure of a sufficient supply, and it is the machine-tender's business to regulate it with the two gates by permitting the surplus to return into the chest.

Every corner of the box in which stuff might lodge, is to be filled up by a board put diagonally across it.

The permanent flow of stuff from the box back into the chest makes it necessary that it should be located above the top of the latter, and at the same time be easily accessible to the machine-tender.

The machine requires a much more diluted stuff than is furnished by the chest. The pulp is therefore mostly led from the regulating box into a mixing box, where it is thinned out by the addition of fresh water.

A box, or save-all, gathers the water which leaves the pulp while on the wire-cloth, and empties it into a fan-pump on the driving side of the machine.

73. Fan-Pump and Mixing Box.—The fan-pump throws the received liquid up into the mixing box, thus saving not only fresh water, but also colouring, sizing matters, and fibres which may have escaped from the pulp with water.

The arms or wings revolve fast within a circular casing of the same shape as the fan, so that the whole space will be constantly scoured. The stuff enters through the centre on the side not pierced by the shaft, is pushed forward by the wings into an outlet at the periphery, and can be forced through a pipe to any desired height.

A large cast-iron or metal casing, extending over the fan-pump and opening into it, receives the water from the save-all.

Instead of using a separate mixing box, the pan, above the fan-pump, is frequently made large enough to serve in its place. In that case the stuff flows from the regulating box directly into the fan-pump receiver, where it is diluted with the water from the save-all, and a fresh supply from a water-pipe. By this arrangement a separate mixing box is not only saved, but the stuff and water in passing through the fan-pump together are more thoroughly mixed.

The violent beating of the fan-pump sometimes increases the froth on heavily-sized pulp, and a separate mixing box may for this reason be preferred for sized papers.

The speed of the fan-pump must be regulated, so that it will forward all the liquid which is poured into it.

74. **Sand-Tables, Pulp-Dressers, and Apron.**—Rags, even if they are carefully sorted, carry with them sand, coal, iron, and other heavy impurities, the weight of which will cause them to separate from the diluted pulp, if it is spread over a sufficiently large surface.

Whenever the paper is found to be sandy, it is a sure indication that the sand-grates in the engines or the sand-tables, or both, are either insufficient or not well attended to.

Any location and form will answer for sand-tables, provided they be large enough, and thereby permit the pulp to flow slowly over them.

If they consisted of only one compartment, the deposits would be carried along to the end of the box, where they would accumulate, and ultimately escape with the pulp. They must therefore be divided into numerous small divisions by means of low gates or weirs. These gates are placed square across the tables in such a way that the deposits cannot be carried over them. Provision must also be made for a quick discharge of these deposits when the tables are to be cleaned out.

The cheapest and most common sand-tables consist of flat wooden boxes with partition boards, which slide in or out between strips fastened to the sides.

A very good table is made of sheet zinc bent up and down, so as to form a succession of bags. Such zinc tables can be taken out bodily, to be emptied and washed.

From the sand-tables the pulp flows into strainers, and should be spread on them as uniformly as possible.

75. Strainers.—The stuff always contains bundles of fibres which have escaped the knives of the beaters by lodging in some corner, or strings, which have been formed of separated fibres by the agitator in the stuff-chest, or on the passage to the pulp-dresser. Light impurities, such as rubber, wood, paper, or straw, are suspended in the stuff, besides the heavy ones which failed to be deposited in the sand-grates and tables.

Every particle of such matters makes a spot in the paper, and it is therefore very important that they should be kept out.

The pulp dressers, strainers, or knotters, guard the entrance to the paper-machine, and should not let anything pass but well-prepared and separated fibre.

76. Bar-Screens.—One of the oldest screens consists of a brass box, the extended arms of which receive an up and down movement by knockers in the usual way. The bottom of the box is formed of narrow brass bars, shaped like the common grate-bars of a steam-boiler furnace.

These bars are pierced by two light brass rods, bearing rings of thin sheet copper or brass between the bars. The thickness of the metal of these rings determines the openings between the bars, and can be varied by means of sets of different sizes. Both ends of the bars fit into cavities in the sides of the screen-box, and form a solid bottom with as many openings or slits as there are bars.

The openings in these screens cannot be made narrow or fine enough for the better grades of paper, and if they are taken apart, it is very difficult to put them together again in such a manner that the distance between the bars will be everywhere the same. The slightest increase of width between two bars allows the knots to pass and makes the screen useless. These strainers, though costly, are very substantial, require no renewal of plates, and answer very well for lower grades of paper or a preliminary screening.

77. **Plate-Screens.**—The screens most generally used consist of brass or composition plates, about 10 by 30 inches, and $\frac{2}{16}$ to $\frac{3}{8}$ inch thick, with narrow parallel openings cut into them by a machine in such a way that they are finest on the upper side, which carries the pulp, and widen out gradually below. The pulp, which has once passed the fine openings on top, thus flows freely through the constantly enlarging channels.

The passage of pulp through the slits wears off the metal and widens the openings. To avoid frequent renewals, the hardest possible composition, which has yet elasticity enough to bear the constant knocks without breaking, should be used for these plates.

Some kinds of stuff are so *free*, that the water escapes quickly through the slits, and leaves the fibres on the screen. Whenever this is the case, the liquid in the vat must be, contrary to the general rule, kept as high as the plates, so that the fibres are held afloat until they pass through the openings.

The knocker-wheels are exposed to much wear and tear, and should always be chilled. A chilled-iron knocker-wheel outlasts many common soft iron ones.

The cast-iron frames, which carry the vat, must be bolted on a very solid foundation. If they stand loosely, the power, which is intended to shake the screens, will vent itself on the whole structure, and the motion of the screens is lessened in the same proportion.

Wooden screen-frames wear out and render frequent repairs necessary, while iron or brass ones give little or no trouble.

78. **Ibotson's Strainer.**—Ibotson has invented a combination of strainers, by which difficulties are overcome.

The pulp enters the screens through two inlets, and flows over the plates, passing over every part of them. Both screens are put in motion by knockers in the usual way, and the pulp which passes through the slits, proceeds over the lip directly to the wire of the paper-machine. All of that portion of the pulp which could not pass through the slits descends at the end of the course, through an outlet-trough, into the auxiliary strainer. The stuff, which makes its way through the plates of this strainer, is pumped back again into the upper screens, and the knots and impurities remaining on it are taken out by hand in the ordinary manner.

79. Suction Strainers.—In some mills pulp-dressers are used, which act entirely by suction without vibration.

The screens are stationary, and form a perfect diaphragm across the vat, so that nothing can pass in anywhere, except through the openings in the plates. The bottom of the vat consists of a rubber-lined plate, which is moved up and down by means of levers. Its action is exactly like that of a pump, and it forces the pulp through the openings by suction only.

The continued motion soon wears out the rubber-joints, and causes frequent and costly renewals.

Screens of this kind, working by suction only, cannot be efficient for any but the finest and best prepared pulp, or such as has previously passed another screen, as the knots are not prevented from settling in the slits, and may obstruct the passage of the pulp.

80. Revolving Screens.—Many revolving screens have been constructed and again abandoned, but those built by George Bertram, of Edinburgh, seem to have found favour with paper-makers in England and Scotland.

The two strainers revolve in vats filled with pulp, impelled by gearing, and their four sides are covered with ordinary screen-plates.

Their interior is provided with a rubber suction arrangement or bellows, moved from the centre, like the piston of a pump; the pulp is thus drawn in from the outside through the slits of the plates, and discharged into the troughs, which conduct it to the wire of the machine.

The knots, which were unable to pass through the openings, remain in the vats, and would soon fill them up, so that the strainers would have to be stopped for the purpose of cleaning them, if they could not be drawn off into the auxiliary screen, which is of ordinary construction.

81. **Reversed Screens**, in which the stuff passes through the plates from below, are sometimes used to give a last cleaning to pulp which has previously passed other pulp-dressers. The knots and other heavy parts fall to the bottom instead of adhering to the plates and obstructing them; but if there should be too many impurities, the vat would have to be cleaned out so often, that the loss of time and pulp would become of more importance than the usefulness of the screens in improving the quality of the paper.

82. **Disposition, Size, and Management of Strainers.**—Mr. Planche recommends the use of three different strainers, in the following order :—

1st. A strainer of the ordinary vibrating kind.

2nd. A pulp-dresser worked by means of a suction plate.

3rd. A strainer in which the pulp passes through the plates from below.

The first one of these screens has the widest openings and retains the coarsest impurities, while each succeeding one is cut finer, and keeps out knots which have passed through the preceding ones.

It is a subject of controversy whether the stuff will be better cleaned by such a succession of screens than in passing only once through finely-cut plates, of which there are enough provided to permit of its being spread over a large surface.

The danger that the knots, which may pass through

some faulty or worn-out opening, may reach the wire-cloth and enter into the paper, is certainly greater in the latter case, but the former system is more complicated.

The plates may be cut finer as their number is increased.

The addition of a second strainer of different construction, to correct the faults of the first set, would in most cases be repaid by an improvement in the quality of the paper.

It is the machine-tender's duty to regulate the speed and vibrations of the screens, as well as to remove the thick stuff and impurities on the plates of ordinary pulp-dressers. If this is not done, the passage of the stuff will be obstructed, and it must run over the top. The screenings should be taken out at regular intervals, and without hammering or knocking the plates, because any such rough action would defeat the object of the pulp-dresser, and force through the openings those knots and impurities which are to be kept out.

83. Connection of the Screen-Vat with the Apron.—It is necessary to have a gate or valve between the pulp-dresser and the wire, by which the flow of pulp can be suddenly stopped or started at will.

84. Aprons.—To support the apron, brackets are fastened to the posts which carry the breast-roll. A solid brass plate, bordered by a flange about 2 inches high all around except on the side against the wire, is usually bolted to these brackets. The open gap between this plate and the wire is bridged over by a piece of leather or oil-cloth, one end of which is fastened to the edge of the plate by means of screws, while the other end rests on the moving wire. It must reach far enough to lay flat on the wire, and it is turned up on the sides to prevent the pulp from escaping in that direction.

If oil-cloth is used, a strong piece of canvas must be placed under it, as otherwise the friction of the wire on the end which rests there would soon tear it. This constant friction wears out even the best aprons, and makes frequent renewals necessary.

Whenever the width of the sheet is to be changed, the turned-up sides must be undone, shifted, and readjusted at the proper places.

85. **The Wire-Cloth and its Attachments.**—The wire is the part of the machine on which the paper is made; it represents the "Form" of the paper-makers of old. It is woven on a loom similar to those used for cotton and linen goods, on which brass wire is substituted for yarn. When the required length of cloth is finished, it is taken from the loom and the ends are sewed together by hand, so that it will form an endless wire-cloth.

The qualities which constitute a good wire-cloth are:

That it is uniformly woven, or that the threads are parallel with, and at equal distances from, each other.

That the wire thread be tough, pliable, equally thick in every part, and capable of suffering a strong tension without tearing apart or breaking. The comparative strength of the wire threads can easily be tested by trying how much weight can be held suspended by pieces of the same length without breaking.

That the seam should be made with great care, so that

it will neither make too distinct a mark on the paper, nor break any sooner than the cloth itself.

Annealed wires are softer and more pliable than ordinary ones, and it is claimed that they will last longer under the same circumstances.

The number of a wire-cloth represents the number of threads contained in one inch of its length; No. 60, the number most used, contains 60 threads in 1 inch. The finer qualities of paper are mostly made on No. 70, with 70 threads in 1 inch. As a rule, the higher numbers are used for fine, and the lower ones for coarse paper.

It is very important that all the rolls, especially the large ones, should be strong enough to endure without springing the heavy pressure to which they are sometimes subjected by a tightly-stretched wire.

It is the tendency of our time to increase constantly the width and speed of the machines, and yet we find these large and fast machines sometimes furnished with rolls and shafts of the same diameter as those for slow and narrow ones. If the rolls are not strong and stiff enough they will bend; the wire will be stretched more in one place than another, causing it to run unevenly and wear out fast. Copper or brass stretch-rolls of five inches diameter, may be sufficiently strong for a 62-inch wire, while they should not be of less than 7 or 8 in. diameter for an 86-inch wire.

86. **The Couch-Rolls** especially should have a copper casing of not less than $\frac{5}{16}$ or $\frac{3}{8}$ inch thickness, and a diameter of from 12 to 15 inches for wide machines, supported inside by many iron spiders on a strong iron shaft.

Hand-made paper was taken from the mould, and stretched out on a felt by a clever movement, and the workman who performed this part of the operation was called the "coucher." The couch-rolls between which the paper passes after it has been formed on the wire, are intended to do his work and press out water besides.

They are for this purpose mostly covered with wool jackets or endless pipe-shaped web, which, to fit close, is used of rather small size, and expanded by a stretcher to the proper width.

Two wooden keys, forming a parallelogram by joining on the diagonal, are often driven and hammered into the jackets, and left there until they are sufficiently widened.

It is of importance that the upper couch-roll should be put on the lower one exactly parallel, after a new wire has been put on the machine. In fact, all the rolls must be parallel, or the wire will not run well; but if the upper couch-roll is out of its true line, the wire will become twisted and run in wrinkles, or stretch unevenly, and spoil in a very short time. The journals of the couch-roll run in boxes fastened to levers, supported by and turning on pivots or studs, so that the roll can be lifted off and on quickly, and without changing its parallel position, by simply turning these levers up or down.

The action of the suction-boxes in addition to their weight causes the ultramarine to settle to the lower side of the paper, and to counteract this influence the upper coucher has been pierced with numerous holes, and connected with a pump which draws the air out of it, thus creating suction on the upper side; but we have not been able to learn anything about the practical working of this plan.

The lower coucher is driven by a pulley, and can be put in and out of motion by a coupling and lever.

87. Tube-Rolls.—It is evidently of importance that we should be enabled to draw as much water as possible from the web while it is on the wire. If we observe the discharge of water through a wire-cloth, we find that more water leaves where it is supported by one of the tube-rolls, or at the points of contact, than in the open spaces. This is probably due to the stream which constantly flows over the surface of these rolls, connecting with the pulp on the wire, and thus drawing the water from it by contact or capillary attraction, while its weight alone has to force it away where there are no rolls.

It is therefore advisable to use a great number of tube-rolls.

88. Suction-Boxes.—Much water is also extracted by the suction-boxes. They are water-tight boxes made of wood or metal, which have no communication with the air except through the wire which is passing over them. The water is drawn from these boxes by suction, produced in the oldest machines by bell-shaped air-pumps. At the present time syphons or suction-pumps are used for that purpose.

If very wide paper is made and the slides are near the ends, the screws moving them project considerably outside of the box, and form an obstruction to the free access and passage to and along the machine.

When a suction-box refuses to work, it will mostly be found that air has gained admittance somewhere; sometimes one side, or only a corner, does not touch the wire.

We have seen much trouble from such simple causes, when, by raising the low corner or side by means of a key, perfect suction would be obtained.

89. **Dandy-Roll.**—Two suction-boxes, at a short distance from each other, are supplied to most machines. Between them, if used at all, on top of the wire and supported by bearings, fastened on the bars, is a wire-roll, open at both ends.

It is constructed of a hollow brass shaft with numerous brass spiders, which carry wooden or sheet-copper strips of the length of the roll. Strong wire is wound around the frame, thus formed, in numerous circles, and the wire-cloth is sewed to it. This is the dandy-roll, justly named so, because it gives to the paper any desired fashionable appearance. It closes up the web a little tighter, and covers, like good cloth on a worthless body, some of its defects. If no particular mark or wove-paper is wanted, the dandy may be covered with the same wire-cloth as that on which it runs. But if any water-mark or impression, such as laid, or wove paper, a name or figures are desired, wires representing every line of the design must be sewed on the cloth covering the dandy. If the design is to fill just one sheet, the circumference of the roll must be exactly as long as a sheet.

The projecting wires press into the already formed but yet soft paper, and displace some pulp. The paper is therefore thinner in those places, and shows the lines through their greater transparency. With a skilfully covered dandy, water-marks of almost any pattern can be produced.

The friction received from the contact with the wire-cloth turns the roll, but if the latter is too heavy, parts of the wet sheet will adhere to it so closely that they are torn out. The dandy is therefore made as light as possible, and no weights or screws are used on it; for the same reason it is impracticable to make it of very large diameter. The bearings can, however, be set higher or lower by means of set-screws, thus regulating, to some extent, the pressure of the roll on the paper.

The stands carry a wooden strip, to which a piece of felting is tacked. This felting touches the dandy all along; it serves as a doctor, and retains particles of pulp which may be carried up from the web.

90. **Save-All and Water-Pipes.**—Between the suction-boxes and the breast-roll, close under the numerous little tube-rolls, extends the save-all, supported by stands on both sides. It is a flat wooden box, about 3 inches deep, receiving all the liquid which leaves the pulp above it, and emptying it into the fan-pump. The outlet from the save-all to the fan-pump is closed by a simple gate, and can be opened by pulling the handle on the front side of the machine.

To the save-all is also attached on its lower side a wooden doctor-board, which constantly scrapes the breast-roll, and is therefore covered with felt on the touching edge. A water-pipe pours in a little stream of water, by which it is kept clean.

As soon as the wire has left the couch-roll on its return trip, it is washed by a steady stream from a shower-pipe. This is done for the purpose of removing any particles of pulp which may stick to the wire, and the shower will be as much more effective as the pressure of water, which can be brought to bear upon it, is stronger. The pipe should at all events be supplied from the highest available reservoir. If this washing is not thoroughly done, the pulp which adheres to the wire will be removed by the next roll which it meets, and wind itself round it.

Wherever this is the case, mostly on the ends, the diameter of the roll is increased by the coat of pulp; the wire is thereby stretched, soon bulged out in all its length, and marks or breaks the paper. To correct these misfortunes the machine-tender, to whose inattention they must be ascribed, resorts to stretching the whole wire-cloth to the same length to which the bulged-out portions have been extended.

91. **Stretch-Roll.**—This he does principally with the stretch-roll, and he should take care to move the boxes on both sides to exactly the same distance from the bars, so that the roll will remain level.

Each time the threads of the wire-cloth are extended in this way, they are also weakened, and while even the length of a well-managed cloth increases through continued tension in the course of time, it speaks badly for the quality of either the machine, wire, or machine-tender, if it is violently lengthened by stretching. Many wires are not allowed to die a natural death from regular wear and tear, but are killed by repeated stretching—the result of the described causes, or of ignorance or neglect.

The much-used doctor will be found an efficient remedy against the collection of pulp on the ends or any other part of the rolls, and especially the stretch-roll.

92. **Stuff-Catchers.**—A box, extending under the coucher

and the shower-pipe receives all the pulp washed off from the wire by the latter. Whenever the machine stops or starts, some of the pulp, too thin to form a sheet, goes into this box, and it is here where the largest and most valuable portion of wasted stuff is gathered. It can, with difficulty only, be removed from there while the machine is running. It is preferable to let the contents empty themselves through a spout, starting from the bottom of the box, and leading to a stuff-catcher situated anywhere near or below the machine, so that the stuff can flow into it.

Sometimes all the pulp and liquid lost on the machine is gathered in a tub or cistern, and pumped into an upper story, whence it is drawn off to be used in place of the pure water, with which the pulp is usually emptied from the beaters.

Both receiving tubs, the one below the machine and the one above the beaters, must be furnished with agitators to prevent the floating fibres from settling on the bottom.

93. **The Shaking Motion.**—The two posts and the frames are respectively connected by iron cross-bars, so that both side-frames may be considered a unit as far as the shaking motion is concerned. The post on the back or driving side of the machine has usually an elongation upwards, and attached to its highest end is the connecting rod, which communicates the shaking movement to the machine. It is placed high enough to allow a man to pass under it. A solid cast-iron, rather ornamental, column, called the shake-post, contains an upright shaft, driven at the lower end by bevel wheels, which in their turn are moved by a pulley on the horizontal shaft.

Bevel friction-wheels are sometimes used instead of the bevel cog-wheels, but as the friction on their surfaces may be reduced by grease, water, or condensed steam, their action is not so reliable as that of cog-wheels.

At its upper end the shaft carries a fly-wheel, which can be turned by hand and serves to start the shake.

It is necessary that the speed as well as the length of the shake-movement may be quickly changed. A set of pulleys of different sizes, or cone-pulleys, give the different speeds, and the eccentricity, which can be easily changed by simply sliding the upright journal to one or the other side, determines the length of the shake.

It is this shaking movement, though it is very trifling (about ¼ inch), which makes the Fourdrinier paper superior to that made on a cylinder-machine. While without it, the fibres would only be laid parallel in the direction in which the wire runs, this cross motion makes them intertwine themselves in all directions, so that they become felted as completely as hand-made paper.

It is evident that the structure and strength of the paper depends to some extent upon this movement, but only experience can teach what speed and length of shake should be used for a certain kind of paper.

The shake must be regulated so that as much water as possible will have escaped through the wire before the paper reaches the suction-boxes; but, as the fibres will only felt themselves as long as they are suspended in water, it must also be prevented from leaving too soon.

94. **The Deckels.**—The width of the paper on the wire is limited by a pair of deckels on both sides. These

deckels are always set a few inches farther apart than the width of the trimmed paper, to allow a margin for shrinkage on the dryers and for trimming.

The deckels are about $1\frac{1}{2}$ inch square endless straps of vulcanised rubber, which have taken the place of those sewed together of cloth, as formerly used. They run over a number of flanged brass pulleys, two of which are fastened to a movable arm. By turning this arm up or down, and then fastening it with the set-screw, the deckels can be stretched to fit close on the rolls.

To prevent any sharp bends of the deckels, which might cause them to crack and wear out, the end-rolls must be large—the larger the better. The rolls nearest to the suction-boxes are fastened on to a separate shaft and stands, but all the other ones are supported by the brass frames, which are themselves carried on the shafts. These shafts are hollow brass tubes, the ends of which are filled with solid wrought-iron bodies. Iron screws, driven by a worm and worm-wheel, fit into these ends, and extend some distance into the brass tubes, which are slotted out on their lower side to an equal distance. A brass nut, sitting on the screw inside, connects through this lower opening with a concentric collar or hub, which forms part of the frame. The screw, being stationary, moves the nut, and with it the deckel-frame in accordance with the turns of the worm-wheel, as directed by the crank. The smaller ends of the tubes which carry the brass-cased screws rest in ball and socket bearings, in which they can change their position. If these bearings were straight and the journals immovable, the shaking motion would exercise a constant strain on the tube, with a tendency to spring or bend it. The two end-rolls, or rather drums, are wide enough to allow the deckels to slide on them as far as the deckel-frames can be shifted.

The friction of the wire is sufficient, without any driving-pulley, to communicate to the deckels its own speed, and, since it is indispensable that both should proceed together, this answers the purpose perfectly.

The frames are stationary and not allowed to touch the deckels anywhere, as they might arrest their movement, and it is left entirely to the weight and tension of the deckels to make them lay flat and close on the wire.

The deckels must not be allowed to have any more play than necessary between the flanges of the carrying pulleys, as they might move sideways to and fro, and give an uneven edge to the paper.

95. The Gates.—As soon as the stuff has left the apron it reaches the gates, consisting of brass sheets, which extends across the wire between the deckels, and are fastened to the frames. Since their length must be variable to suit the different widths of paper, they are made of two pieces, sliding against each other, and fastened together by bolts. Their height above the wire can be regulated by a screw at each end; and to secure a sheet of uniform thickness, this must be done with great care. The paper cannot be of even thickness all across unless the lower edges of these gates are in every point equally distant from the wire.

As it is not desirable that a sheet should begin to form itself before the pulp has reached the gate, the leather or cloth extension of the apron must cover the wire to within about one inch from it. The wire in constantly pulling this leather, gradually stretches it forward until it sometimes reaches the gate, obstructing perhaps in one or more spots the passage under it. The paper becomes thin and weak in such places, and breaks.

The 3 to 6 inches space between the two gates are filled to some height with pulp, which receives here the strongest shaking motion, and being well diluted, intertwines the fibres more thoroughly than on any other part of the wire.

96. Length of the Wire-Cloth.—The paper-makers of the United States seem to have settled on 33 feet as the most appropriate length of a wire-cloth; but if the speed is going on increasing at the rate it has done lately, the pulp will not remain sufficiently long on such a wire to form a sheet, and longer ones will become a necessity.

97. Wire-Guides.—It has been mentioned that the wire is kept in its place by the guide-roll, which is regulated by the machine-tender. Many mechanisms have been invented to do this automatically, and some are used successfully.

98. Press-Rolls and Housing.—The paper after it has left the wire is fully formed, and efforts are next directed to free it from water. This is done by means of two or more presses, which are alike in principle.

The first, or wet press, consists of a pair of rolls, of not less than 12 to 15 inches diameter, the journals of which rest in upright cast-iron stands on the frames of the machine. The lower roll is connected by a coupling with a driving shaft on the back or driving side of the machine, and receives motion directly, while the upper one is turned by friction.

Both rolls are of cast-iron and mostly hollow, and though the upper one may be very heavy, its weight alone is hardly sufficient to exercise the necessary pressure. Sometimes screws are brought to bear directly on the journals of the upper roll. The pressure may thereby be considerably increased; but if any solid substance happens to get between the rolls, something must give way and break.

If levers and weights are used, the upper roll can be lifted up by the intruder, which passes through without breaking any part of the machinery.

Of whatever size or material the rolls may be, they must be made of greater diameter in the centre than at the ends, in order to press equally at all points. If made perfectly straight or of uniform diameter, they will spring open in the middle and form a hollow place.

It is of the greatest importance that the two rolls should touch each other in a straight line through their entire length when the proper amount of pressure is put upon them, because the paper cannot lose as much water in a hollow place as in the other parts, therefore remains wet, is consequently weaker in that particular spot, and breaks on its subsequent journey towards the reel. A certain quantity of water is evaporated from every part of the paper on the dryers; the spot which has not been sufficiently pressed, and contains the most water, will therefore remain moist when the balance of the web is dry.

The couch and press-rolls are usually made a trifle full—sometimes $\frac{1}{16}$ to $\frac{1}{32}$ inch thicker in the middle than on the ends—and then require a pressure which is at all times strong enough to bend them into straight lines, and frequently stronger than would otherwise be necessary.

The weight, which must be brought to bear on the presses, depends not only on the quality and thickness of the paper, on the dilution of the pulp, and its treatment on the wire, but especially on the condition of the felts. While the felt is new and clean the water can pass through it easily, but a strong pressure is required after the pores have become partially filled up.

99. **Brass and Rubber-Cased Press Rolls.**—The surface of an iron roll will rust when not in use; the rust is transferred to the felt as soon as the rolls start, and from it to the paper. The surface of the roll may be entirely smooth when new, but the rust will soon make it rough.

To avoid this and to secure a permanently bright and close-grained surface, the iron rolls are frequently covered with a brass casing about $\frac{3}{8}$ inch thick.

100. **Doctors.**—The upper press-roll is always supplied with *a doctor*, which prevents parts of the paper or the whole sheet, when broken, from going all round and thickening on the roll.

These doctors consist of a cast-iron body, somewhat longer than the roll, with journals on the ends, which rest in bearings bolted to the stands.

A thin steel, brass, or hard-rubber plate is secured on the body all along, resting on the roll and scraping it. The levers fastened on the journals carry weights on their ends, and increase the pressure of the doctor-blades against the roll. Hard-rubber blades are stiff and hard enough for the purpose, while they do not cut the roll as much as metal.

It has been observed that the doctor-blades, if allowed

to remain in the same position all the time, will cause the roll to wear unevenly into hills and hollows, and to obviate this a slow vibrating motion is given to them.

101. The Felts.—The paper is led through the first press in the same direction in which it moves on the wire.

The upper press-roll makes the side of the paper with which it is in direct contact more compact and smooth than the lower side, which rests on the felt; and in order to obtain a similar surface on both sides, it must be reversed in the second press; the side which was in contact with the felt in the first press must be brought in contact with the metal surface of the roll in the second press.

For this purpose the wet felt carries the paper underneath the second press and several feet beyond it to a point where it is taken off by hand, led over the two paper-rolls and laid on the second felt. This second or press-felt, moving in a direction opposite to that of the pulp on the wire, carries the paper backward through the second press. It will adhere to the upper press-roll and turn with it upwards to a point where it is taken off again, and led over another paper-roll placed above the press-rolls, to the drying cylinders.

102. Felt and Paper-Carrying Rolls.—Two distinct sets of rolls carry the felts and paper respectively.

The felt-rolls are sometimes subjected to a very heavy strain by tightly-stretched felts, while the paper-rolls have but little strain upon them.

The wider the machine is, or the longer the rolls are,

the easier will they be bent; and their diameter should therefore increase with the width of the machine.

When a felt is to be changed, the upper roll is raised several inches by means of the hand-wheel, the front-side journal of the lower-roll is then lifted from its bearing by means of a lever or jack-screw, the old felt is drawn out over the roll between its front journal and bearing, and the new one is passed in through the same narrow opening. This can however not be done before all those carrying-rolls, which are situated inside of the felt, are removed.

103. **Wet and Press-Felts.**—All the water which is pressed out of the paper must pass through the felts, and, as the first press necessarily takes out a great deal more than the second press, the wet-felt must permit the passage of a larger quantity of water than the press-felt. The wet-felts are therefore of a light and open web, while the press-felts are thicker and heavier.

It is well known that every kind of woollen cloth, especially felts, shrink considerably when wet, the lower grades more so than the finer ones. They should therefore be always made from 6 to 8 inches wider than the press-rolls, and the frames should be far enough apart, and the felt-rolls long enough, to give them plenty of room to spread.

104. **Spread and Stretch-Rolls.**—To counteract the tendency to contract or shrink, one or more of the felt-rolls are covered with spirals or worms. These worms are made of two strips of heavy felting, about 1 to $1\frac{1}{2}$ inches wide, starting from a point in the middle of the roll,

where they are fastened, and winding in spiral or screw lines around it, leaving at least two inches distance between the turns of the strips, until they reach the ends, where they are fastened again. To hold these strips on iron-rolls, they are sewed together in the middle, from where they start, and tied to the roll there and at the ends with strings.

Each felt is provided with one or two guide-rolls, by means of which it is kept in its place. If, for instance, a felt moves to the front side of the machine, the machine-tender advances the journal of the guide-roll on that side in the direction in which the felt runs over it, until it goes back. If the felt shifts to the back side, the guide-roll journal on the front side is screwed back against its line of travel, until it remains permanently in the middle.

The constant wear and tear weakens and elongates the felts, and it becomes often necessary to preserve their stiffness by stretching.

The wet-felt being the longer, most exposed, and weaker web, should have plenty of stretch-room, or its screws should be as long as possible.

105. Felt-Washers.—The stuff, especially if short, well loaded with clay, or heavily-sized, will soon fill up the pores of the felts, so as to prevent the passage of water through them. To wash the wet-felt, a shower-pipe is placed over it on the return trip, and immediately after having been soaked with water, it receives the friction beating of two wooden wings, fastened on a horizontal shaft below the felt, which is revolved with high speed by a small pulley on its back end. This washing operation,

though it takes only a few minutes, necessitates stoppage of the machine, and various attempts have been made to have the felt washed all the time while it is running.

The second, or press-felt, does not require to be washed so frequently as the wet-felt, and nothing is therefore provided for this purpose. It is simply soaked with water, and receives a beating with sticks, in the hands of the machine-tender.

By this rough proceeding it is frequently spoiled, and many paper-makers prefer therefore to change the press-felt whenever it becomes stiff or filled up, and to clean it outside of the machine in a felt-washer.

106. **Air-Roll.**—Sometimes there is air between the paper and the felt, which, as it cannot pass through the press, forms bubbles right before the rolls, and bulges the paper out. To prevent this, a copper-tube roll rests on the paper, and, pressing it by its weight, makes it impossible for the air to advance any farther.

107. **Clutch.**—Each press should be supplied with a clutch and lever, by which it can be either stopped or started at will. When, for example, one of the felts is to be washed, the press must run while the rest of the machine may stand still.

108. **Management of Felts.**—It is of the greatest importance that the felt-rolls, as well as all other rolls, should be level, square, and parallel with each other. If only one of them is out of line, the felt may become wrinkled, and, passing through the press in that way, will be cut and spoiled.

109. Taking the Paper through the Presses.—The paper is taken from the wire by hand, and laid on the wet felt, the first carrying roll of which should be as near to the lower coucher as it can be placed, without obstructing the passage of paper and sometimes of thick pulp to the box underneath—not over two inches distant. It is, again by hand, taken from the wet-felt to the press-felt, and from the upper roll of the second press to the dryers, as already described.

A narrow passage for the convenience of the machine-tender is usually left between the second press and the dryers, and a plank, resting on the frames and bridging the felt and paper between the first press and the stretch-roll of the press-felt, provides another passage to the driving side.

110. Drying-Cylinders.—The first drying-cylinders were made of copper, but they have been almost altogether superseded by cast-iron ones, as the latter are less expensive, less liable to have their surfaces damaged, and will not change temperature as quickly as copper ones. In cases where iron dryers cannot be used, the brass or copper shell should be very heavy, in order to combine as fully as possible the advantages of both.

It is not only necessary that the surface of a dryer should be a perfect cylinder, but also that the body should be thoroughly balanced.

111. Drying.—The process of making paper by hand is closely imitated by the machine up to the dryers. Hand-made paper is, however, not dried artificially, but simply exposed to the air, and contracts to $\frac{9}{10}$ or even $\frac{4}{5}$ of its size during that operation. The fibres join each other closely in all directions, and produce a firm and tough sheet.

On the machine the paper can only shrink very little in the direction in which it runs, because it is constantly drawn out and stretched. Sometimes the original sheet, formed on the wire, is even lengthened out through the action of the presses. Nothing prevents the paper from shrinking in width or at right angles to the line in which it travels; but it is usually dried so fast that it has no time to contract, and the sheet on leaving the dryers is found to be very little narrower than it is at the press-rolls.

By subjecting the paper slowly to gradually-increased temperatures, the natural way of drying by air will be imitated as much as possible, and the qualities of hand-made paper, its tough and yet pliable body, will be in a measure given to the product.

If the paper is dried fast and strongly heated, it will acquire qualities the reverse of those we aim at; it will be brittle, of porous appearance, and sometimes even badly sized, although the engineer may have sized the pulp in precisely the same manner which, until then, gave always good results. When the paper is too violently heated it becomes cockled and unfit for use, and must be worked over again.

The larger the heating surface over which the paper passes, the better.

The greatest economy of fuel will be reached when the number of dryers is so large that all the steam used can be permitted to condense in them.

The first cylinder should have the lowest, and the last one the highest, temperature, and the steam-pipe, coming from the generator, should therefore connect with the pipe at the end, where the paper leaves the dryers.

112. Quantity of Fuel Required for Drying Paper.—It has been found by theoretic calculation, as well as from experience, that about $\frac{1}{2}$ pound of coal is required to produce steam enough wherewith to dry 1 pound of paper. The fuel, boilers, pipes, number and arrangement of dryers, &c., modify this quantity; and in some mills, where, on account of insufficient heating surface, high-pressure steam must be allowed to rush through the cylinders without condensing, 1 pound of coal or more is used to dry a pound of paper.

113. Calenders.—The calenders consist mostly of iron rolls, placed in a stack or nest. The lower roll is coupled to a shaft driven by belt and pulleys, and all the others are moved by friction. The surfaces of all the rolls, of whatever diameter they may be, have therefore the same speed.

The machine-tender takes the paper from the last dryer by hand, puts it between the uppermost pair of rolls, and guides it all through the stack. The object of this operation is to compress the web, and especially its surface, so that the pores or hollow spaces between the fibres will be filled up, and the whole mass solidified.

Paper, like other materials polished in a similar manner, acquires thereby a smooth, glossy appearance.

To produce a uniform sheet, it is necessary that the rolls should fit perfectly on one another, and that their surfaces be true and smooth.

If the light shines through between the rolls while they are standing or running empty, it is an evidence that they are not true; if their surface is rough, they cannot be expected to give a smooth one to the paper; and in neither case are they fit for the duty allotted to them. Their polishing power is proportionate to their weight or pressure; they are therefore made heavy, and weights on a leverage, or screws, are brought to bear on the upper roll.

Whenever the paper passes between two rolls, it is subjected to the polishing pressure; the greater therefore the number of rolls, the more effective is the stack. On the other hand, very small rolls have little weight, and cannot exercise the same pressure as large ones; but their smaller and sharper circles act more acutely upon the paper.

The continued friction of the paper, especially of common qualities, wears out common iron rolls in an incredibly short time. It gradually wears an opening as wide and as thick as the sheets, while the ends through which the paper never passes remain intact and keep the rolls at the original distance. When the calenders have reached that condition they are of no use, and must be taken out and turned or ground off. If made of pretty hard iron, and kept in good order, they may give satisfaction, but if they are soft, and must be frequently turned off at a heavy expense, it will be found more advantageous to dispense with them altogether.

114. **Chilled Rolls.**—Chilled-rolls, with a hard steel-like surface, are now justly taking the place of common iron ones.

They make a better surface because they are harder, and last for years without being redressed or ground. They are cast in thick iron moulds called chills. The contact of the molten iron with the cold chill produces a change of texture in the surface of the roll, turning the iron into a close-grained, fine metal, resembling steel. This transformation is most perfect at the surface, decreases towards the centre, and leaves the main body unaffected. A change of texture may be observed in such rolls as far in as one inch from the circumference.

115. Steaming the Paper.—The paper, coming fresh from the dryers, is rather hard, and cannot take the impression of the calenders as easily as if its surface were somewhat humid. The dry paper is therefore frequently moistened with steam before it passes through the calenders. An iron steam pipe about $\frac{1}{2}$ inch wide, with numerous holes as large as pin-heads, which might be called a steam shower-pipe, is fastened to the frames of the calenders, a few inches below the sheet, where it first enters. As soon as the valve is opened, little jets of steam start all along from the holes in the pipe, striking and moistening the paper. If it should be found desirable, two of these pipes may be used, one on each side of the calenders, to moisten both sides of the paper.

The steam from these shower-pipes strikes the rolls and makes them wet, whenever the paper breaks and leaves the calenders uncovered.

116. Reels.—The paper, before it is allowed to go to the cutter, is invariably wound up on reels. The cutter, being supplied from full or finished reels only, can be stopped or started while the machine is running, and is to a large extent independent of it. It is also desirable that the cutters should not run too fast, and this is attained by cutting the paper from two or three reels, that is, two or three thicknesses of paper at once.

To do this the machine must be supplied with at least three or four reels, two or three of which feed the cutter, while one receives the paper. The reels are fastened either on stationary or revolving frames. The stationary frames hold three reels perpendicularly, one above the other.

117. Trimming and Cutting. Slitters.—The edges of the paper which has been wound up on the reels, after it has passed the dryers, are always rough, and must be trimmed. The sheets ordered by the consumers are seldom large enough to require the whole width of the paper, and the web is therefore cut lengthways into two or three parts by means of small circular knives called slitters.

118. Cutters.—The action of the cutting-machine is an imitation of that of the knife guided by hand.

The table, on which the paper is spread for the operation of the latter, is represented in the former by a stationary bed-knife; and while the sheets are passing over or resting on it, the movable knife begins to cut at one end, and gradually proceeds all across, like scissors.

The cutters may, according to their construction, be classed as continuous feed-cutters and stop-cutters.

On continuous feed-cutters, of which we know only one kind, the paper is moved forward without any check; it travels on while it is being cut off.

The reels deliver the paper continuously also to the stop-cutter, as long as it is in motion, but the web is prevented from proceeding between the knives; it stops while the cut is being made

119. **Paper in Endless Rolls.**—Paper is for many purposes required in rolls instead of sheets; for instance, for hanging, roofing, bag paper, and for some printing presses. In such cases it is conducted through the feed-rolls of the cutter, all other parts of which are at rest, or if no cutter is supplied to the machine, directly from the slitters to a shaft, which is turned by friction like the driving shaft of a reel, and wound up on it.

If it were rolled directly around the shaft, the paper-rolls would fit on it so tightly that they could not afterwards easily be removed. The shaft has therefore a key-seat all along, and into it is placed a thick wooden or iron key-shaped rod, which projects considerably above the surface, and has a handle by which it can be pulled out. The paper is thus forced to form a larger circle on the shaft than its circumference, and the rolls can easily be slipped off as soon as the key is withdrawn.

It has been proposed not to cut the lower qualities of paper, such as wrapping, &c., into sheets, but to punch rows of little holes across the lines where they are to be separated.

They could thus be easily torn apart, like postage-stamps, and might be wound up and supplied in rolls, saving thereby the expense of labour and material for counting, folding, and packing.

120. **Motive-Power.**—The paper-machine must always be driven by a motor of its own, which furnishes power for it, the stuff-pump, and stuff-chest only.

It is of the greatest importance that the speed of the machine should be regular, as every part of it is set and arranged for the production of a certain quantity of paper

per minute. The valves which admit water to the mixing-box, steam to the dryers, and draw water from the suction-boxes, are opened for a certain speed. The relative speeds of the different parts of the machine are also adjusted for it, and as all this cannot be quickly altered with every change of movement produced by an irregular motor, the paper will either break or show defects of some kind.

If the engines, rag-cutters, rotaries, or super-calenders were driven by the same motor as the machine, the stopping or starting of one of them would reduce or increase the amount of power which moves the machine, and cause a corresponding increase or reduction of speed. Even the best of governors cannot regulate promptly enough to avoid a serious derangement.

If water power is the motor, its independence should extend even to the water-supply. If one forebay is used for the wheels which drive the mill, as well as the machine, the starting of the mill, or of a beater only, will cause a sudden demand for more water, which may lower for a short time the head common to both; the machine wheel will lose power, and its speed become reduced. The water-wheel which drives the machine should have a forebay and penstock, independent of any other one, and be fed directly from the race.

If the mill has an insufficient water-power, the paper-machine should be driven by a separate steam-engine; the engine must be selected with a view to regular speed, and that a quick-acting, reliable governor is indispensable.

High-pressure engines are the most suitable, because the paper-machine furnishes in the drying cylinders an excellent condenser. The steam from the boilers is to be conducted to the engine through as large and short pipes as possible, rather too large than too small, so that nearly the full pressure may be available in the cylinder. After the steam has acted in the engine, and transmitted the larger part of its power to the piston, it escapes at a low pressure through short and capacious pipes into the drying cylinders, in which the exhausted steam is utilised.

The general experience is, that it takes but little more fuel to drive the machine and dry the paper with one stream of live steam, than for drying the paper only.

The following universally accepted theory gives a scientific explanation for this fact :—

Boiling water and steam of the pressure of our atmosphere have both a temperature of 212 degrees Fahrenheit, and nevertheless considerable time and the continued application of heat are required for the transformation of the former into the latter. The heat which has been expended during this time on the water, though it has not raised its temperature one degree and is apparently lost, is treasured up by it and is called "latent heat."

Before boiling water can change its liquid state into that of steam, it must have taken up a large amount of this latent heat.

121. **Size and Speed.**—It is natural that a manufacturer should endeavour to make as much paper with his machine as possible, because very little more capital and labour are required for the production of 100 feet than of 50 feet per minute, if he has power and machinery enough to do it.

The faster the machine runs, the less time is allowed for the formation of the paper and for the escape of the water on the wire, presses, and dryers, and if the speed is increased too much the quality must suffer.

www.ingramcontent.com/pod-product-compliance
Lightning Source LLC
Chambersburg PA
CBHW030317170426
43202CB00009B/1046